Transgenic Modification of Germline and Somatic Cells

Transgenic Modification of Germline and Somatic Cells

Edited by

R. B. Flavell

Director, John Innes Institute
Norwich
and
AFRC Institute of Plant Science Research
University of Sussex
UK

and

R. B. Heap

Director, AFRC Babraham Institute
Cambridge
UK

A discussion meeting published on behalf of The Royal Society by Chapman & Hall

The Royal Society

CHAPMAN & HALL
London · Glasgow · New York · Tokyo · Melbourne · Madras

Published by Chapman & Hall, 2–6 Boundary Row, London SE1 8HN

Chapman & Hall, 2–6 Boundary Row, London SE1 8HN, UK

Blackie Academic & Professional, Wester Cleddens Road, Bishopbriggs, Glasgow G64 2NZ, UK

Chapman & Hall Inc., 29 West 35th Street, New York NY1001, USA

Chapman & Hall Japan, Thomson Publishing Japan, Hirakawacho Nemoto Building, 6F, 1–7–11 Hirakawa-cho, Chiyoda-ku, Tokyo 102, Japan

Chapman & Hall Australia, Thomas Nelson Australia, 102 Dodds Street, South Melbourne, Victoria 3205, Australia

Chapman & Hall, India, R. Seshadri, 32 Second Main Road, CIT East, Madras 600 035, India

First edition 1993

Printed in Great Britain by The Alden Press, Osney Mead, Oxford

ISBN 0 412 55510 7

A catalogue record for this book is available from the British Library

Library of Congress Cataloging-in-Publication data

Contents

Contributors

N. D. Allen,	Department of Molecular Embryology, AFRC Institute of Animal Physiology and Genetics Research, Babraham, Cambridge CB2 4AT, UK.
M. Antoniou,	Laboratory of Gene Structure & Expression, National Institute for Medical Research, The Ridgeway, Mill Hill, London NW7 1AA, UK.
A. L. Archibald,	AFRC Institute of Animal Physiology and Genetics Research, Edinburgh Research Station, Roslin, Midlothian EH25 9PS, UK.
G. D. Austin,	Monsanto Company, 700 Chesterfield Village Parkway, St Louis, Missouri 63198, USA.
S. C. Barton,	Wellcome/CRC Institute of Cancer and Developmental Biology, University of Cambridge, Tennis Court Road, Cambridge CB2 1QR, UK. *and* Department of Molecular Embryology, AFRC Institute of Animal Physiology and Genetics Research, Babraham, Cambridge CB2 4AT, UK.
C. Bernelot Moens,	Samuel Lunenfeld Research Institute, Mount Sinai Hospital, 600 University Avenue, Toronto, Canada M5G 1X5. *and* Department of Molecular and Medical Genetics, University of Toronto, Toronto, Canada.
M. Berry,	Laboratory of Gene Structure & Expression, National Institute for Medical Research, The Ridgeway, Mill Hill, London NW7 1AA, UK.
E. de Boer,	Laboratory of Gene Structure & Expression, National Institute for Medical Research, The Ridgeway, Mill Hill, London NW7 1AA, UK.
D. Bradley,	Department of Genetics, AFRC Institute of Plant Science Research, John Innes Centre, Colney Lane, Norwich NR4 7UH, UK.
R. Carpenter,	Department of Genetics, AFRC Institute of Plant Science Research, John Innes Centre, Colney Lane, Norwich NR4 7UH, UK.
C. T. Caskey,	Howard Hughes Medical Institute *and* Institute for Molecular Genetics, Baylor College of Medicine, One Baylor Plaza, Houston, Texas 77030, USA.
A. J. Clark,	AFRC Institute of Animal Physiology and Genetics Research, Edinburgh Research Station, Roslin, Midlothian EH25 9PS, UK.
P. R. Clemens,	Institute for Molecular Genetics, Baylor College of Medicine, One Baylor Plaza, Houston, Texas 77030, USA.
E. Coen,	Department of Genetics, AFRC Institute of Plant Science Research, John Innes Centre, Colney Lane, Norwich NR4 7UH, UK.

L. Copsey, Department of Genetics, AFRC Institute of Plant Science Research, John Innes Centre, Colney Lane, Norwich NR4 7UH, UK.

H. G. Dickenson, Department of Plant Sciences, University of Oxford, South Parks Road, Oxford OX1 3RB, UK.

N. Dillon, Laboratory of Gene Structure & Expression, National Institute for Medical Research, The Ridgeway, Mill Hill, London NW7 1AA, UK.

S. Doyle, Department of Genetics, AFRC Institute of Plant Science Research, John Innes Centre, Colney Lane, Norwich NR4 7UH, UK.

R. Elliott, Department of Genetics, AFRC Institute of Plant Science Research, John Innes Centre, Colney Lane, Norwich NR4 7UH, UK.

J. Ellis, Laboratory of Gene Structure & Expression, National Institute for Medical Research, The Ridgeway, Mill Hill, London NW7 1AA, UK.

C. J. Farr, Laboratory of Human Molecular Genetics, Imperial Cancer Research Fund, Lincoln's Inn Fields, London WC2A 3PX, UK.

A. C. Ferguson-Smith, Wellcome/CRC Institute of Cancer and Developmental Biology, University of Cambridge, Tennis Court Road, Cambridge CB2 1QR, UK.
and
Department of Molecular Embryology, AFRC Institute of Animal Physiology and Genetics Research, Babraham, Cambridge CB2 4AT, UK.

P. Fraser, Laboratory of Gene Structure & Expression, National Institute for Medical Research, The Ridgeway, Mill Hill, London NW7 1AA, UK.

M. E. Fromm, Monsanto Company, Plant Sciences Division, 700 Chesterfield Village Parkway, St Louis, Missouri 63198, USA.

P. N. Goodfellow, Laboratory of Human Molecular Genetics, Imperial Cancer Research Fund, Lincoln's Inn Fields, London WC2A 3PX, UK.

F. Grosveld, Laboratory of Gene Structure & Expression, National Institute for Medical Research, The Ridgeway, Mill Hill, London NW7 1AA, UK.

O. Hanscombe, Laboratory of Gene Structure & Expression, National Institute for Medical Research, The Ridgeway, Mill Hill, London NW7 1AA, UK.

S. Hantke, Department of Genetics, AFRC Institute of Plant Science Research, John Innes Centre, Colney Lane, Norwich NR4 7UH, UK.

J. Hurst, Laboratory of Gene Structure & Expression, National Institute for Medical Research, The Ridgeway, Mill Hill, London NW7 1AA, UK.

A. Imam, Laboratory of Gene Structure & Expression, National Institute for Medical Research, The Ridgeway, Mill Hill, London NW7 1AA, UK.

P. A. Jones, Kenneth Norris Jr Comprehensive Cancer Center, University of Southern California, 1441 Eastlake Avenue, Los Angeles, California 90033, USA.

R. Jorgensen, Department of Environmental Horticulture, University of California, Davis, California 95616–8587, USA.

M. Lindenbaum, Laboratory of Gene Structure & Expression, National Institute for Medical Research, The Ridgeway, Mill Hill, London NW7 1AA, UK.

R. Lovell-Badge, Laboratory of Eukaryotic Molecular Genetics, MRC National Institute for Medical Research, The Ridgeway, Mill Hill, London NW7 1AA, UK.

D. Luo, Department of Genetics, AFRC Institute of Plant Science Research, John Innes Centre, Colney Lane, Norwich NR4 7UH, UK.

M. McClenaghan, AFRC Institute of Animal Physiology and Genetics Research, Edinburgh Research Station, Roslin, Midlothian EH25 9PS, UK.

P. McSteen, Department of Genetics, AFRC Institute of Plant Science Research, John Innes Centre, Colney Lane, Norwich NR4 7UH, UK.

K. Mitani, Howard Hughes Medical Institute, Baylor College of Medicine, One Baylor Plaza, Houston, Texas 77030, USA.

M. Mooney, Department of Genetics, AFRC Institute of Plant Science Research, John Innes Centre, Colney Lane, Norwich NR4 7UH, UK.

A. B. Moseley, Institute for Molecular Genetics, Baylor College of Medicine, One Baylor Plaza, Houston, Texas 77030, USA.

A. Nagy, Department of Molecular and Medical Genetics, University of Toronto, Toronto, Canada.

F. J. Perlak, Monsanto Company, Plant Sciences Division, 700 Chesterfield Village Parkway, St Louis, Missouri 63198, USA.

S. Philipsen, Laboratory of Gene Structure & Expression, National Institute for Medical Research, The Ridgeway, Mill Hill, London NW7 1AA, UK.

S. B. Prusiner, Departments of Neurology and of Biochemistry and Biophysics, University of California, San Francisco, California 94143, USA.

S. Pruzina, Laboratory of Gene Structure & Expression, National Institute for Medical Research, The Ridgeway, Mill Hill, London NW7 1AA, UK.

S. Raguz-Bolognesi, Laboratory of Gene Structure & Expression, National Institute for Medical Research, The Ridgeway, Mill Hill, London NW7 1AA, UK.

W. Reik, Department of Molecular Embryology, AFRC Institute of Animal Physiology and Genetics Research, Babraham, Cambridge CB2 4AT, UK.

C. Robinson, Department of Genetics, AFRC Institute of Plant Science Research, John Innes Centre, Colney Lane, Norwich NR4 7UH, UK.

J. Romero, Department of Genetics, AFRC Institute of Plant Science Research, John Innes Centre, Colney Lane, Norwich NR4 7UH, UK.

J. Rossant, Samuel Lunenfeld Research Institute, Mount Sinai Hospital, 600 University Avenue, Toronto, Canada M5G 1XS.
and
Department of Molecular and Medical Genetics, University of Toronto, Toronto, Canada.

H. Sasaki, Wellcome/CRC Institute of Cancer and Developmental Biology, University of Cambridge, Tennis Court Road, Cambridge CB2 1QR, UK.
and
Department of Molecular Embryology, AFRC Institute of Animal Physiology and Genetics Research, Babraham, Cambridge CB2 4AT, UK.

L. Sefton,	Laboratory of Human Molecular Genetics, Imperial Cancer Research Fund, Lincoln's Inn Fields, London WC2A 3PX, UK.
R. Simon,	Department of Genetics, AFRC Institute of Plant Science Research, John Innes Centre, Colney Lane, Norwich NR4 7UH, UK.
J. P. Simons,	AFRC Institute of Animal Physiology and Genetics Research, Edinburgh Research Station, Roslin, Midlothian EH25 9PS, UK.
C. R. Somerville,	DOE Plant Research Laboratory, Michigan State University, East Lansing, Michigan 48824, USA.
D. M. Stark,	Monsanto Company, Plant Sciences Division, 700 Chesterfield Village Parkway, St Louis, Missouri 63198, USA.
J. Strouboulis,	Laboratory of Gene Structure & Expression, National Institute for Medical Research, The Ridgeway, Mill Hill, London NW7 1AA, UK.
M. A. Surani,	Wellcome/CRC Institute of Cancer and Developmental Biology, University of Cambridge, Tennis Court Road, Cambridge CB2 1QR, UK. *and* Department of Molecular Embryology, AFRC Institute of Animal Physiology and Genetics Research, Babraham, Cambridge CB2 4AT, UK.
D. Talbot,	Laboratory of Gene Structure & Expression. National Institute for Medical Research, The Ridgeway, Mill Hill, London NW7 1AA, UK.
R. Wallace,	AFRC Institute of Animal Physiology and Genetics Research, Edinburgh Research Station, Roslin, Midlothian EH25 9PS, UK.
C. B. A. Whitelaw,	AFRC Institute of Animal Physiology and Genetics Research, Edinburgh Research Station, Roslin, Midlothian EH25 9PS, UK.

Preface

Research in animals and plants is frequently 'departmentalized' and funded according to Kingdoms by granting bodies. The use of transgenes to address biological questions in all Kingdoms led us to propose to the Royal Society that fundamental and strategic studies in animals and plants involving transgenes should be presented in one meeting, rather than, as so often happens, in different scientific societies. The two-day Discussion Meeting held in July 1992, and reported here provided insights into how transgenes are being exploited to discover new knowledge in animals and plants. The papers were presented by leading investigators in the biological sciences, and the book reflects an experiment in interdisciplinarity which was declared a successful venture by the large crowd of participants and delegates.

The transgenic area is one of high scientific interest and sporadic, yet intense biotechnological euphoria. This is dramatically illustrated among the following papers which show how genetic maps of animals and plants produce new knowledge of disease incidence in humans, and how the design of transgenes can result in biodegradable plastic in higher plants, human pharmaceutical proteins in livestock, or bacterial proteins in cotton crops to protect against insect damage.

Spectacular insights have been gained into the control of sex determination and early development. The juxtaposition papers on the differential programming of certain animal genes according to their parental origin alongside the occurrence of gene imprints in plants, the dissection of gene switching systems that determine the type of haemoglobin in animals alongside the control of flowers and bracts in plants, revealed unexpected surprises for those whose scientific curiosity had not consciously crossed the Kingdom boundary in these areas. The power of genome modification is also reflected in the ability to delete genes from embryonic stem cells to study their function by reverse genetics, to transduce cells and correct for single gene disorders in humans, and to produce models of disease as in the case of neurodegenerative conditions of animals and humans. The meeting stimulated great interest among the media who shared constructively in the novelty of the programme and helped to reach a wider audience. Space limitations, however, did not permit us to incorporate the excellent poster presentations of many colleagues.

We are indebted to the Royal Society for the opportunity to field such a lively meeting, our colleagues for the excellence of their presentations, and to staff of the Royal Society, London, for the professional execution of the arrangements. It is for the reader to judge whether the success of the experiment is conveyed by the scientific prose!

R. B. Flavell
R. B. Heap

1

Genetic maps

P. N. Goodfellow, L. Sefton and C. J. Farr

SUMMARY

The same genetic principles are common to all animals, plants and microorganisms. One consequence of the generality of genetic principles is that methods developed in one organism can often be used in other species. Recent advances in mammalian molecular genetics, genome analysis and gene mapping have been found to be directly applicable to other animals and even plants. This is facilitating the rapid construction of detailed genetic maps in a wide variety of species. Similar advances are being made in the development of manipulation techniques, which allow the genomic deletion and addition of specific genes as well as more subtle alterations. Armed with detailed maps and precise manipulation techniques it is possible to associate genes with phenotypes, test the function of genes and to unravel complex traits that depend on the interaction of different genes.

1. INTRODUCTION

Improvement of plant crops by altering yield, disease resistance and growth range has been a goal since the beginning of plant cultivation. Similarly improvements in fecundity, rates of growth and disease tolerance have been sought in animal husbandry. These agricultural aims have been approached traditionally by plant and animal breeding methods which are rooted in statistical analysis (Falconer 1960). The low resolution of the available genetic maps prevented the use of more precise methods of genetic analysis (Thoday 1961; Tanksley *et al.* 1989). The marriage of traditional genetics with molecular biology to produce molecular genetics has changed this outlook in three ways: first, by exploiting DNA sequence variation directly as a source of genetic markers, it is now relatively simple to construct a high resolution genetic map in any species; second, the map position of an identified locus can be used as the starting point for cloning any gene, a process variously called 'reverse genetics' (Ruddle 1984) or 'positional cloning' (Collins 1992), and third, the cloned gene can be used for genome modification by transgenesis.

Techniques for map creation and genome modification developed in one species are often directly applicable to another species. In closely related species, such as mammals, the maps themselves may be sufficiently similar to be applicable between species (O'Brien & Graves 1991; Searle *et al.* 1989). Even in distantly related organisms, such as fruit flies, mammals and flowering plants, both regulatory genes and structural genes may be sufficiently conserved that cloned genes from one organism can be used to recognize genes with similar biochemical functions in the other organisms (Frigerio *et al.* 1986; Burri *et al.* 1989).

Although the cloning of genes which cause major inherited diseases has been spectacular in the past ten years (Kunkel *et al.* 1985; Rommens *et al.* 1989), application of the same methods to either rare inherited diseases or multifactorial diseases is limited by both available family material and the major effort involved in cloning by positional cloning methods. The human genome project aims to provide a high resolution linkage map, identification and mapping of every gene and, eventually, the sequence of the genome (Dulbecco 1986). In the future, this will obviate much of the work needed in any specific study aimed at the analysis of any particular phenotype. Many of the techniques developed for genome mapping and analysis in humans have found a ready application in animal genetics and an echo in plant genetics.

In agricultural species, maps of the genome facilitate the identification of genes and this information can be exploited in breeding, however, even greater benefits can be expected if genes are cloned and then used for directed genome modification. The addition of genes by transgenesis and the subtle modification of genomes by homologous recombination will allow the rapid construction of plant and animal varieties designed specifically for different purposes and environments.

This review is intended to provide a brief summary of the use of molecular techniques in the construction of genetic maps and to provide a background for the following papers in this volume.

† Present address: Department of Genetics, University of Cambridge, Downing Street, Cambridge CB2 3EH, U.K.

2. POLYGENES AND QUANTITATIVE TRAIT LOCI

Many complex phenotypes are under the control of different genes acting together; these genes have been termed polygenes. In humans, several of the major diseases afflicting our society have complex inheritance patterns pointing to the existence of polygenes and environmental factors. In agricultural species, disease resistance, yield, quality and environmental tolerance are all subject to polygenic inheritance. Many of these traits are quantitative and the loci to which the corresponding polygenes map have been named quantitative trait loci or QTLs.

The contribution of any one 'polygene' to a specific phenotype may be small and dependent on environmental conditions. The identification of polygenes has been difficult because of the lack of suitable markers that could be used to track them in crosses (Thoday 1961; Tanksley *et al.* 1989). Moreover, the few markers that were available were often morphologically based and may have contributed to the phenotype under study. A number of techniques were developed to identify polygenes; for example, the construction of congenic lines by repeated backcrossing combined with selection for graft rejection was used to identify major and minor histocompatibility loci and recombinant inbred lines produced by inbreeding the progeny of an F2 generation were used to analyse drug resistance (Taylor 1976). These methods are laborious and tend to identify only those genes that make a large contribution to the phenotype. In practice, plant and animal breeding has been based on statistical methods with no attempt to identify the specific genes involved. This means that information gained in one cross cannot be applied to any future cross designed to yield further enhancement. High resolution linkage maps, composed of markers that are phenotypically neutral and that can be scored in any cross, provide the raw material for the identification of all the genes that contribute to a phenotype.

3. GENETIC MARKERS AND LINKAGE MAPS

Linkage maps are the fundamental resource for genetic analysis: they allow the association of genes with phenotypes. The ideal map would be comprised of a set of highly polymorphic markers spread evenly throughout the genome. The major problem in map creation is the number of markers that can be scored in any one cross. The major problem in map exploitation is the probability that a marker is polymorphic in a test cross. Sequence variation at the DNA level provides an unlimited source of polymorphic markers for map creation and exploitation. The three most commonly used sequence-based markers are:

1. Restriction fragment length polymorphisms (RFLPs). These markers are based on Southern blotting and simple sequence changes leading to the loss and gain of restriction sites (Solomon & Bodmer 1979; Botstein *et al.* 1980). The probes used include cDNA clones and random genomic fragments. The disadvantages of these markers are their limited polymorphism and the labour-intensive methods needed to identify and score them.

2. Minisatellites and microsatellites. These consist of a variable number of tandem repeats (VNTRS) and are frequently highly polymorphic with many different alleles at each locus. The core of the repeat sequence in many minisatellites is related to the sequence GGCAG-GAXG (Jeffreys *et al.* 1985). The repeat units in microsatellites are mono-, di-, tri- and tetranucleotides. In humans, minisatellites tend to cluster towards the telomeric ends of chromosomes, whereas, microsatellites are apparently more evenly distributed (Royle *et al.* 1990). A further advantage of microsatellites is that they can be scored by the polymerase chain reaction (PCR). The best studied class of minisatellites are the CA dinucleotide repeats. With an estimated 12000 polymorphic $(dC-dA)_n \cdot (dG-dT)_n$ loci in the human genome, these markers alone would yield a genetic map with an average resolution of 0.3–0.5 cM (Weber 1990). Microsatellites are found universally in eukaryotes (Tautz & Renz 1984) and their high levels of polymorphism make them the markers of choice for many purposes. The disadvantage of microsatellite markers is the expense of isolating large numbers.

3. Random amplified polymorphic DNA (RAPDS). A single, short, oligonucleotide of random sequence is used as a primer in each PCR reaction (Williams *et al.* 1990). This is an inexpensive method for generating large numbers of markers very quickly, however, the patterns produced are strain specific and it is often not possible to recognize pairs of alleles in a cross. A further disadvantage is the difficulty in accessing the genome at the locus corresponding to the defined marker as this requires further cloning and sequencing. Nevertheless, for many purposes, especially in plant genetics, RAPDS are the markers of choice (Waugh & Powell 1992).

There are many other methods for detecting sequence variation some of which can be used for identifying or scoring polymorphism: allele specific hybridization with oligonucleotides (ASO, Wallace *et al.* 1979); single strand conformation polymorphisms (SSCP, Orita *et al.* 1989); denaturing gradient gel electrophoresis (DGGE, Myers *et al.* 1985); chemical cleavage at mismatches (Cotton *et al.* 1988) and two-dimensional gel electrophoresis (Hatada *et al.* 1991).

The possibility of generating unlimited numbers of highly polymorphic genetic markers by DNA based methods has changed genetics and map construction is occuring at a rapid rate in many different plant and animal species. We have chosen a few examples to illustrate the progress and problems in the creation of linkage maps.

(a) *The human map*

Constructing linkage maps of the human genome poses special problems. Humans are an outbred species with a long life cycle, random mating behaviour and small families. Although, Bayesian statistical methods can be used to obviate some of the worst aspects of random mating and small families (Ott

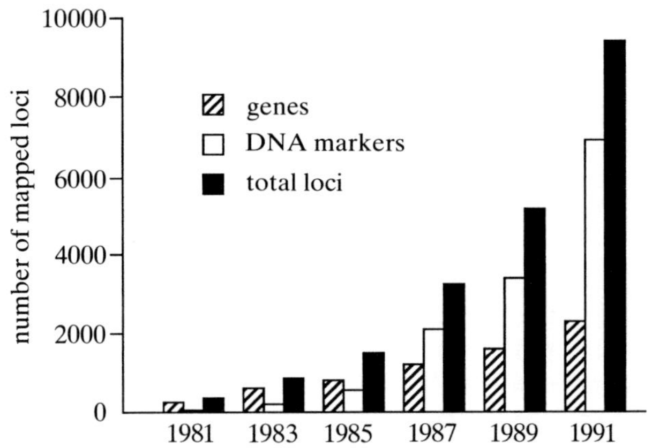

Figure 1. Progress in mapping the human genome. Legend data taken from the Human Gene Mapping Workshop (Human Gene Mapping 1991).

1974), these are poor circumstances for creating maps. Nevertheless, the biggest barrier to map creation for most of this century was the lack of phenotypic traits suitable for use as genetic markers. The impact of molecular techniques on human genetics can be seen in figure 1. The first human autosomal genes were not mapped to a specific chromosome until 1969 (Donahue *et al.* 1968); ten years later 360 markers had been placed on the human gene map by somatic cell genetic techniques (Shows & McAlpine 1979). However, the chance of finding linkage between a disease locus and a known polymorphic marker was negligible. The introduction of molecular probes resulted in a dramatic increase in the number of mapped loci. In 1992, 2325 genes have been mapped, 10000 loci defined by DNA probes and framework linkage maps for the whole genome have been published (Human Gene Mapping 11 1991).

In the creation of maps of the human genome, linkage studies have been extensively supplemented with information gained from somatic cell genetics and cytogenetics. For many years, the only realistic option for mapping a gene to a specific human chromosome was to use human–rodent hybrids. Subchromosomal localization by this approach was restricted by the availability of appropriate translocation chromosomes. Some of the limitations of somatic cell methods have been overcome by radiation hybrid mapping (also known as irradiation and fusion gene transfer). Radiation hybrids are generated by the lethal irradiation of a donor cell line (usually a human–hamster hybrid containing a single human chromosome) and subsequent rescue by fusion to a second cell line (Goss & Harris 1975; Cox *et al.* 1990). The hybrids produced retain fragments of the donor chromosomes including the human chromosome. Marker order and inter-marker distance can be deduced from the frequency of retention of two human markers within the same cell line (Lawrence *et al.* 1991). A potential advantage of this method is that the frequency of chromosome breakage is related to the radiation dose and it is possible to create high resolution maps of chromosomes in regions of low recombination.

New techniques in cytogenetics are also contributing to genome mapping. Fluorescence *in situ* hybridization (FISH) is a sensitive technique which allows the simultaneous sublocalization and ordering of several DNA sequences. The resolution of FISH depends on the degree of condensation of the target chromatin: metaphase chromosomes are tightly packed and permit a resolution of 1 Mb; with the use of less condensed chromatin from interphase nuclei or pronuclei, the resolution may be increased to 50 kb (Brandriff *et al.* 1991).

The human map has been exploited to locate many of the genes causing common single-gene defect human diseases (e.g. Murray *et al.* 1982; Gusella *et al.* 1983) and this has led to the direct benefits of prenatal and presymptomatic diagnosis. The map location has also been exploited to clone several of these genes including those causing muscular dystrophies (Kunkel *et al.* 1985; Brook *et al.* 1992), cystic fibrosis (Rommens *et al.* 1989) and the fragile-X syndrome (Verkerk *et al.* 1991). Nevertheless, the difficulty of this approach can be gauged by the frustrations in cloning the Huntington disease gene which was among the first to be mapped using molecular markers. The increasing saturation of the human map has resulted in the identification of genes causing disease by co-localization with plausible candidate genes encoding proteins of known biochemical function (the 'candidate gene' method, e.g. Marfan syndrome, Dietz *et al.* 1991) and this type of serendipity will become increasingly common as the Genome Project proceeds.

One of the major problems in constructing maps is integration of those constructed by different groups using different techniques. Map integration can only be achieved if groups cooperate by using common sets of markers (reference markers). The construction of maps of the human genome has been aided by two cooperative organisations: the Human Gene Mapping Workshops and the Centre d'Etude du Polymorphisme Humaine (CEPH). Human Gene Mapping Workshops have provided a forum for collaboration on concensus map construction (HGM11 1991). Recently, these meetings have undergone a metamorphosis under the sponsorship of HUGO (Human Genome Organisation) to the Chromosome Co-ordinating Committee Meetings. CEPH maintains and distributes DNA from reference families allowing different groups to type the same families. The immortalised cell lines from the reference families are a renewable source of DNA and the nearest human equivalent to recombinant inbred strains.

(*b*) *The mouse map*

As a model for creating genetic maps of agricultural species, the mouse is a much more appropriate model than humans. The laboratory mouse is an experimental animal with a short life span and a wealth of phenotypic markers many of which affect coat colour and form. In consequence, the mouse linkage map was, for many years, much more advanced than the human map. Even in the mouse, however, DNA markers have been needed for the creation of high resolution maps of general utility.

The most productive approach for creating linkage maps in the mouse has been the interspecific backcross between inbred laboratory mice and inbred strains of *Mus spretus* (Copeland & Jenkins 1991). The use of inbred strains reduces the number of alleles segregating in the cross from four to two and helps the unambiguous definition of the origin of any allele. The use of genetically diverse strains or subspecies increases the probability of detecting sequence variation at any locus. Potential problems in exploiting crosses between genetically divergent strains include suppression or enhancement of recombination due to chromosomal rearrangements, however, with a few minor exceptions these problems have not been observed. Nearly 3000 markers have been placed on the mouse linkage map and a map with an average resolution of 1 cM should be reached soon (Chapman *et al.* 1991).

Although the interspecific cross is now the workhorse of mouse gene mapping, recombinant inbred strains and congenic strains are still used extensively. Recombinant inbred strains are constructed by crossing two inbred strains followed by repeated brother sister mating of individuals of the F2 generation. Once homozygosity has been achieved, individual meiotic events from the F2 mice and subsequent rounds of inbreeding are fixed and these mice provide a renewable resource of meiotic events for map construction (Dietrich *et al.* 1992). Recombinant inbred strains also have a significant advantage for analysing quantitative traits because each genotype is present in all animals in the inbred line. This allows accurate measurement of phenotypic variability for each genotype (Taylor 1976; Lander & Botstein 1989).

Several groups have used high resolution maps to identify the genes contributing to polygenic diseases in the mouse. Genetic crosses between non-obese diabetic (NOD) mice and disease-resistant strains have identified 4–5 loci outside the MHC contributing to diabetes in the mouse (Todd *et al.* 1991; Prochazka *et al.* 1991; Cornall *et al.* 1991; Garchon *et al.* 1991). A similar analysis of hypertension has been performed in the rat (Hilbert *et al.* 1991).

Comparison of the human and mouse maps has revealed a large degree of conservation in map order despite the genetic distance between rodents and primates. On average the blocks of synteny are about 10 cM for autosomes (Searle 1989). The conserved blocks of synteny can be used for predicting homologous genetic diseases and can be helpful for suggesting candidate genes for human diseases. In a recent example, a gene causing Waardenburg syndrome (WS1) was mapped to a region of the long arm of chromosome 2 that is homologous to the proximal part of mouse chromosome 1. Waardenburg syndrome is responsible for over 2% of adult deafness and the tissues affected by this disease are related by their embryonic origin in the neural crest. The mouse mutation *Splotch* maps to the proximal part of chromosome 1 and the mutant mice are characterized by dysgenesis of neural crest cell derived tissues. As the mutation in *Splotch* is due to a deletion in *Pax-3* (Epstein *et al.* 1991) the human homologue HuP2 (Burri *et al.* 1989) was investigated and mutations in this gene were identified in patients with Waardenburg syndrome (Baldwin *et al.* 1992; Tassabehji *et al.* 1992).

(c) *Maps in other mammals*

There are 19 recognized orders of mammals that include 4000 species. Gene maps of 28 species have been published representing eight orders (O'Brien & Graves 1991). The high degree of syntenic conservation observed between mammalian species has allowed the human and mouse gene maps to be used as references for general gene map construction. The large number of cloned human and mouse sequences has provided a ready source of DNA markers.

For breeders of domestic animal species the availability of large numbers of genetic markers offers the possibility of mapping the genes that determine production traits and of using this information to increase the speed and accuracy of selection. One advantage which animal breeders have is that phenotypic information on thousands of animals has been carefully recorded over the years for use in classical biometrical breeding programmes, allowing them to design and generate the ideal family material required for mapping studies.

For linkage analysis of cattle, reference families have been constructed in Australia, Texas and Kenya and are being distributed to the bovine mapping community following the CEPH model. The combined use of VNTRS, microsatellites, *in situ* hybridization and of bovine–hamster somatic cell hybrids has resulted in the majority of the bovine genome being included in a low resolution linkage map. The remarkable conservation of mini- and microsatellites within Bovidae will substantially accelerate the construction of genetic maps in sheep and goats (Georges *et al.* 1991). In general it has been found that the extent of human–bovine chromosomal conservation exceeds that of human–mouse (O'Brien & Graves 1991).

Pig chromosomes were first studied at the beginning of this century and the correct chromosome number of 38 was described as early as 1931 over 20 years before man (Krallinger, 1931). The Pig Gene Mapping project (PiGMap) is based on a shared reference population (three generations) produced from crosses between the Chinese Meisham and the European Large White or European Wild Boar and improved European breeds. Recently, remarkable conservation between human and pigs has been reported for dinucleotide repeats (Wintero *et al.* 1992). The polydisperse nature of the porcine karyotype allows the chromosomes to be sorted effectively by FACS analysis, providing a powerful resource for physical mapping and for the construction of chromosome-specific libraries (Archibald & Haley 1992).

(d) *Maps in non-mammalian vertebrates*

Gene maps in non-mammalian vertebrates are underdeveloped despite the evolutionary insights that might acrue from comparative mapping. Comparisons of fish and amphibian gene maps with those of mammals have identified several syntenic associations apparently conserved through more than 400 million years of vertebrate divergence. Expansion of fish gene maps should provide a valuable tool for the reconstruction of chromosome rearrangement events in mammalian lineages. Fish of the genus *Xiphophorus* (platyfishes and swordtails) are ideal for gene mapping because of their almost complete interfertility, the high level of genetic variability and numerous interspecific hybrids have been established, which should allow for rapid expansion of the *Xiphophorus* gene map (Morizot *et al.* 1991).

A preliminary linkage map of the chicken genome has recently been reported based on RFLP analysis of crosses between two lines of white leghorns (Bumstead & Palyga 1992). The chicken has a number of advantages for the construction of linkage maps: the genome is relatively small and the reproductive potential is high. However, the chicken karyotype is complex, with eight pairs of macrochromosomes and 30 pairs of microchromosomes (most of which are cytologically indistinguishable), as well as the sex chromosomes.

(*e*) *Linkage maps of plants*

The large numbers of morphological markers, the existence of ecotypes, the ease of creation of new variants by mutagenesis and the large numbers of offspring obtainable from individual plants facilitated the construction of detailed linkage maps in several plants (e.g. *Arabidopsis thaliana*, and maize). Nevertheless linkage maps constructed with DNA markers have been a major focus of research; access to the genome is offered by each marker and testing of large numbers of markers can be achieved in a single cross. In construction of maps it has also been possible to exploit a large number of other genetic tools including recombinant inbred strains, aneuploid variants that allow chromosomal localization by dosage, congenic strains constructed by introgression of chromosomal fragments from related species and translocation stocks. A list of some of the available plant linkage maps based on molecular markers is presented in table 1.

The detailed linkage maps have been exploited for mapping genes affecting disease resistance (e.g. Landry *et al.* 1987) and development (e.g. Weigel *et al.* 1992) and have been used to study the evolutionary relationships between different plant species (Bonierbale *et al.* 1988). The largest impact, however, has been on the study of complex phenotypes and polygenes. In several experiments it has been possible to define the Mendelian factors underlying quantitative traits affecting fruit size and quality (Paterson *et al.* 1988; 1991; Edwards *et al.* 1987) and water use (Martin *et al.* 1989). The ability to map a polygene depends critically on both the contribution it makes to the phenotype and the density of the linkage map. Once identified further localization may require construction of congenic or near isogenic lines specifically designed to study a particular polygene. In the short term the mapping of polygenes should facilitate manipulation of plant species to produce variants of improved agricultural performance; in the longer term cloning of polygenes and the direct manipulation of plant genomes should greatly improve the efficiency of producing new varieties.

Table 1. *RFLP linkage maps in plants*

plant species	reference
Arabidopsis thalia	Chang *et al.* (1988)
barley (*Hordeum vulgare*)	Heun *et al.* (1991)
lettuce (*Lactuca sativa*)	Landry *et al.* (1987)
maize (*Zea mays*)	Helentjaris (1987)
potato (*Solaneum tuberosum*)	Bonierbale *et al.* (1988)
rice (*Oryza sativa*)	McCouch *et al.* (1988)
soybean (*Glycine max*)	Apuya *et al.* (1988)
tomato (*Lycopersicon esculentum*)	Bernatzky & Tankersly (1986)

4. PHYSICAL MAPS AND CONSTRUCTION OF CONTIGS

Genome analysis has been transformed by the introduction of long-range restriction mapping (Brown & Bird 1986) and by cloning systems based on yeast artificial chromosomes (YACs, Burke *et al.* 1987). Long-range restriction mapping can be used to measure the precise physical distance between loci defined by DNA probes and has been applied to the genomes of prokaryotes as well as eukaryotes. Before the introduction of long-range restriction mapping, physical distances between loci were estimated indirectly by reference to linkage maps or by cloning of the intervening sequences: a feat that was rarely achieved in higher eukaryotes. Long-range restriction mapping utilizes restriction enzymes ('rare-cutters') that cleave rarely in genomic DNA because of the number of bases in their recognition sites. In many cases, rare cutters are also 'methylation-sensitive' and will not cut DNA in the recognition site has been methylated. Cleaved DNA is separated by pulsed field gel electrophoresis (PFGE) or by related methods such as field inversion gel electrophoresis (FIGE); these methods are capable of separating molecules as large as a few megabases (Schwartz & Cantor 1984; Carle & Olson 1984; Orbach *et al.* 1988). The distribution of CpG methylation and the corresponding long range restriction maps provides clues to genome organization. In mammals, methylation occurs only at CpG dinucleotides and is genome wide except for the CpG rich regions (also known as HTF islands; Bird 1987) found at the 5′ ends of many genes; this means that there is a direct relationship between the distribution of genes and the construction of the long range map (Bird & Brown 1986). Methylation at cytosine in plants occurs at CpXpG trinucleotides as well as CpG nucleotides (Gruenbaum *et al.* 1981). The distribution of methylation in plants has been studied less. However, blocks of methylation free genome appear to correspond to gene rich regions (Antequera & Bird 1988; Moore *et al.* 1992).

The fragments of DNA that can be cloned in YACs are sufficiently large that they can integrate directly with linkage maps. In humans, 1 cM is approximately equivalent to 1 Mb of DNA and YACs with inserts this size are being produced by several groups (Larin *et al.* 1991). By comparison, in *Arabidopsis* 1 cM corresponds to only 140 kb (Chang *et al.* 1988). YACs can be used to rapidly isolate the DNA from target region of the genome and have greatly simplified chromosome walking. The major 'bottleneck' in positional cloning experiments is probably now in gene identification rather than chromosome walking.

Contigs are contiguous, overlapping cloned DNA fragments conceptually arranged in the same order as they are found in the genome. A complete contig of an organism's genome would allow high resolution mapping of any DNA based marker and could be used for instant access to any part of the genome. Initial attempts to construct contigs were based on cosmids (Somerville 1989; Stallings *et al.* 1990) using 'fingerprinting' techniques pioneered in studies on *C. elegans* (Coulson *et al.*, 1986); most recent studies have utilised YACs. In model organisms, such as *C. elegans* (Coulson *et al.* 1991) and *Arabidopsis* (Grill & Somerville 1991) near closure of contigs has been achieved.

The construction of a contig, besides providing an important mapping tool, is an essential first step in sequencing of the genome (Sulston *et al.* 1992).

5. CONCLUSIONS

The construction of genetic maps in plants and animals used to be regarded as an esoteric pastime with few practical consequences. With the introduction of molecular techniques this is no longer the case.

We would like to thank all those who took the time to try and educate us, in particular P.N.G. is grateful for conversations with Dr I. Furner, Dr G. Moore and Dr E. Coen. N. Ellis read the manuscript and made many perceptive comments, most of which we ignored.

REFERENCES

Antequera, F. & Bird, A.P. 1988 Unmethylated CpG islands associated with genes in higher plant DNA. *EMBO J.* **7**, 2295–2299.

Apuya, N.R., Frazier, B.L., Keim, P., Roth, E.J. & Lark, K.G. 1988 Restriction fragment length polymorphisms as genetic markers in soybean *Glycine max*. *Theor. appl. Genet.* **75**, 889–901.

Archibald, A. & Haley, C.S. 1992 The pig gene mapping project (PiGMap). *Gnome News* **10**, 26–28.

Baldwin, C.T., Hoth, C.F., Amos, J.A., da-Silva, E.O. & Milunsky, A. 1992 An exonic mutation in the *HuP2* paired domain gene causes Waardenburg's syndrome. *Nature, Lond.* **355**, 637–638.

Bernatzky, R. & Tanksley, S.D. 1986 Towards a saturated linkage map in tomato based on isozymes and random cDNA sequences. *Genetics* **112**, 887–898.

Bird, A.P. 1987 CpG islands as markes in the vertebrate nucleus. *Trends Genet.* **3**, 342–347.

Bonierbale, M.W., Plaisted, R.L. & Tanksley, S.D. 1988 RFLP maps based on a common set of clones reveal modes of chromosomal evolution in potato and tomato. *Genetics* **120**, 1095–1103.

Botstein, D., White, R.L., Skolnick, M.H. & Davis, R.W. 1980 Construction of a genetic linkage map in man using restriction fragment length polymorphisms (RFLPs). *Am. J. hum. Genet.* **32**, 314–331.

Brandriff, B., Gordon, L. & Trask, B. 1991 A new system for high-resolution DNA sequence mapping in interphase pronuclei. *Genomics* **10**, 75–82.

Brook, J.D., McCurrach, M.E., Harley, H.G., Buckler, A.J., Church, D., Aburatani, H., Hunter, K. *et al.* 1992 Molecular basis of myotonic dystrophy: Expansion of a trinucleotide (CTG) repeat at the 3′ end of a transcript encoding a protein kinase family member. *Cell* **68**, 799–808.

Brown, W.R. & Bird, A.P. 1986 Long-range restriction site mapping of mammalian genomic DNA. *Nature, Lond.* **322**, 477–481.

Bumstead, N. & Palyga, J. 1992 A preliminary linkage map of the chicken genome. *Genomics* **13**, 690–697.

Burke, D.T., Carle, G.F. & Olson, M.V. 1987 Cloning of large fragments of exogenous DNA in yeast by means of artificial chromosome vectors. *Science, Wash.* **236**, 806–812.

Burri, M., Tromvoukis, Y., Bopp, D., Frigerio, G. & Noll, M. 1989 Conservation of the paired domain in metazoans and its structure in three isolated human genes. *EMBO J.* **8**, 1183–1190.

Carle, G.F. & Olson, M.V. 1984 Separation of chromosomal DNA molecules from yeast by orthogonal-field-alternation gel electrophoresis. *Nucl. Acids Res.* **12**, 5647–5664.

Chang, C., Bowman, J.L., Dejohn, A.W., Lander, E.S. & Meyerowitz, E.M. 1988 Restriction fragment length polymorphism linkage map for *Arabidopsis thaliana*. *Proc. natn. Acad. Sci. U.S.A.* **85**, 6856–6860.

Chapman, V.M., Klein, J., Nadeau, J.H. & Silver, L.M. 1991 Encyclopedia of the mouse genome. *Mamm. Genome*, **1**, 1–532.

Collins, F.S. 1992 Positional cloning: Let's not call it reverse anymore. *Nature Genetics* **1**, 3–6.

Copeland, N.G. & Jenkins, N.A. 1991 Development and applications of a molecular genetic linkage map of the mouse genome. *Trends Genet.* **7**, 113–118.

Cornall, R.J., Prins, J.-B., Todd, J.A., Pressey, A., De Larato, N.H., Wicker, L.S. & Peterson, L.B. 1991 Type 1 diabetes in mice is linked to the interleukin-1 receptor and *Lsh/Ity/Bcg* genes on chromosome 1. *Nature, Lond.* **353**, 262–265.

Cotton, R.G.H., Rodrigues, N.R. & Campbell, R.D. 1988 Reactivity of cytosine and thymidine in single base-pair mismatches with hydroxylamine and osmium tetroxide and its application to the study of mutations. *Proc. natn. Acad. Sci. U.S.A.* **85**, 4397–4401.

Coulson, A., Kozono, Y., Lutterbach, B., Shownkeen, R., Sulston, J. & Waterston, R. 1991 YACs and the *C. elegans* genome. *BioEssays* **13**, 413–417.

Coulson, A., Sulston, J., Brenner, S. & Karn, J. 1986 Towards a physical map of the genome of the nematode *Caenorhabditis elegans*. *Proc. natn. Acad. Sci. U.S.A.* **83**, 7821–7825.

Cox, D.R., Burmeister, M., Price, E.R., Kim, S. & Myers, R.M. 1990 Radiation Hybrid Mappng: A somatic cell genetic method for constructing high-resolution maps of mammalian chromosomes. *Science, Wash.* **250**, 245–250.

Dietrich, W., Katz, H., Lincoln, S.E., Sin, H.-S., Friedman, J., Dracopoli, N.C. & Lander, E.S. 1992 A genetic map of the mouse suitable for typing intraspecific crosses. *Genetics* **131**, 423–447.

Dietz, H.C., Cutting, G.R., Pyeritz, R.E., Maslen, C.L., Sakai, L.Y., Corson, G.M. & Puffenberger, E.G. *et al.* 1991 Marfan syndrome caused by a recurrent de novo missense mutation in the fibrillin gene. *Nature, Lond.* **352**, 337–339.

Donahue, R.P., Bias, W.B., Renwick, J.H. & McKusick, V.A. 1968 Probable assignment of the Duffy blood group locus to chromosome 1 in man. *Proc. natn. Acad. Sci. U.S.A.* **61**, 949–955.

Dulbecco, R. 1986 A turning point in cancer research: sequencing the human genome. *Science, Wash.* **231**, 1055–1056.

Edwards, M.D., Stuber, C.W. & Wendel, J.F. 1987 Molecular-marker-facilitated investigations of quantitative loci in maize. 1. Numbers, genomic distribution and gene action. *Genetics* **116**, 113–125.

Epstein, D.J., Vekemans, M. & Gross, P. 1991 Splotch (Sp^{2H}), a mutation affecting development of the mouse neural tube, shows a deletion within the paired homeodomain of *Pax-3*. *Cell* **67**, 767–774.

Falconer, D.S. 1960 *Introduction to quantative genetics*. New York: Raven Press.

Frigerio, G., Burri, M., Bopp, D., Baumgartner, S. & Noll, M. 1986 Structure of the segmentation gene paired and the Drosophila PRD gene set as part of a gene network. *Cell* **47**, 735–746.

Garchon, H.-J., Bedossa, P., Eloy, L. & Bach, J.-F. 1991 Identification and mapping to chromosome 1 of a susceptibility locus for periinsulitis in non-obese diabetic mice. *Nature, Lond.* **353**, 260–262.

Georges, M., Gunawardana, A., Threadgill, O.W., Lathrop, M., Olsaker, I., Mishra, A. & Sargeant, L.L. *et al.* 1991 Characterisation of a set of variable number of tandem repeat markers conserved in Bovidae. *Genomics* **11**, 24–32.

Goss, S.J. & Harris, H. 1975 New method for mapping genes in human chromosomes. *Nature, Lond.* **255**, 680–683.

Grill, E. & Somerville, C. 1991 Construction and characterization of a yeast artificial chromosome library of *Arabidopsis* which is suitable for chromosome walking. *Molec. Gen. Genet.* **226**, 484–490.

Gruenbaum, Y., Naveh-Many, T., Cedar, H. & Razin, A. 1981 Sequence specificity of methylation in higher plant DNA. *Nature, Lond.* **292**, 860–862.

Gusella, J.F., Wexler, N.S., Conneally, P.M., Naylor, S.L., Anderson, M.A., Tanzi, R.E. & Watkins, P.C. *et al.* 1983 A polymorphic DNA marker genetically linked to Huntington's disease. *Nature, Lond.* **306**, 234–236.

Hatada, I., Hayashizaki, Y., Hirotsune, S., Komatsubara, H. & Mukai, T. 1991 A genomic scanning method for higher organisms using restriction sites as landmarks. *Proc. natn. Acad. Sci. U.S.A.* **88**, 9523–9527.

Helentjaris, T. 1987 A genetic linkage map for maize based on RFLPs. *Trends Genet.* **3**, 217–221.

Heun, M., Kennedy, A.E., Anderson, J.A., Lapitan, N.L.V., Sorrells, M.E. & Tanksley, S.D. 1991 Construction of a restriction fragment length polymorphism map for barley (*Hordeum vulgare*) *Genome* **34**, 437–447.

Hilbert, P., Lindpaintner, K., Beckmann, J.S., Serikawa, T., Soubrier, F., Dubay, C., Cartwright, P. *et al.* 1991 Chromosomal mapping of two genetic loci associated with blood-pressure regulation in hereditary hypertensive rats. *Nature, Lond.* **353**, 521–529.

Human Gene Mapping 11. 1991 *Cytogenet. Cell Genet.* **56**, 1–2200.

Jeffreys, A.J., Wilson, V. & Thein, S.L. 1985 Hypervariable 'minisatellite' regions in human DNA. *Nature, Lond.* **314**, 67–73.

Krallinger, H.F. 1931 Cytologische Studien in einigen Haussaugetieren. Habilitationsschrift 1988. Arch. F. Tierernahrwig und Tierzucht. Abt. B5 Breslau. (Cited in Gustavsson, I. *Hereditas* **109**, 151–157.

Kunkel, L.M., Monaco, A.P., Middlesworth, W., Ochs, H.D. & Latt, S.A. 1985 Specific cloning of DNA fragments absent from the DNA of a male patient with an X chromosomal deletion. *Proc. natn. Acad. Sci. U.S.A.* **82**, 4778–4782.

Lander, E.S. & Botstein, D. 1989 Mapping Mendelian factors underlying quantitative traits using RFLP linkage maps. *Genetics* **121**, 185–199.

Landry, B.S., Kesseli, R.V., Farrara, B. & Michelmore, R.W. 1987 Genetic map of lettuce (*Lactuca sativa*) with restriction fragment length polymorphism, isozyme, disease resistance and morphological markers. *Genetics* **116**, 331–337.

Larin, Z., Monaco, A.P. & Lehrach, H. 1991 Yeast artificial chromosome libraries containing large inserts from mouse and human DNA. *Proc. natn. Acad. Sci. U.S.A.* **88**, 4123–4127.

Lawrence, S., Morton, N.E. & Cox, D.R. 1991 Radiation hybrid mapping. *Proc. natn. Acad. Sci. U.S.A.* **88**, 7477–7480.

Li, Y.K. & Ponce-de-Leon, F.A. 1992 Partitioning of the chicken genome by microcell hybridisation. *Poult. Sci.* **71**, 151–160.

Ludecke, H.-J., Senger, G., Claussen, U. & Horsthemke, B. 1989 Cloning of defined regions of the human genome by microdissection of banded chromosomes and enzymatic amplification. *Nature, Lond.* **338**, 348–350.

Martin, B., Nienhuis, J., King, G. & Schaefer, A. 1989 Restriction fragment length polymorphisms associated with water use efficiency in tomato. *Science, Wash.* **243**, 1725–1728.

McCouch, S.R., Kochert, G., Yu, Z.H., Wang, Z.Y., Khush, G.S., Coffman, W.R. & Tankersly, S.D. 1988 Molecular mapping of rice chromosomes. *Theor. appl. Genet.* **76**, 815–829.

Moore, G., Cheung, W., Foote, T., Gale, M., Koebner, R., Leitch, A. & Leitch, I. 1992 Key features of cereal genome organisation as revealed by the use of cytosine methylation-sensitive restriction endonucleases. *Genomics* (In the press.)

Morizot, D.C., Slaugenhaupt, S.A., Kallman, D.K. & Chakravarti, A. 1991 Genetic linkage map of fishes of the genus Xiphophorus (Teleostei: Poeciliidae). *Genetics* **127**, 399–410.

Murray, J.M., Davies, K.E., Harper, P.S., Meredith, L., Mueller, C.R. & Williamson, R. 1982 Linkage relationships of a cloned DNA on the short arm of the X chromosome to Duchenne muscular dystrophy. *Nature, Lond.* **300**, 69–70.

Myers, R.M., Fischer, S.G., Lerman, L.S. & Maniatis, T. 1985 Nearly all single base substitutions in DNA fragments joined to a GC-clamp can be detected by denaturing gradient gel electrophoresis. *Nucl. Acids Res.* **13**, 3131–3146.

Nelson, D.L., Ledbetter, S.A., Corbo, L., Victoria, M.F., Ramirez-Solis, R., Webster, T.D., Ledbetter, D.H. & Caskey, C.T. 1989 Alu polymerase chain reaction: a method for rapid isolation of human-specific sequences from complex DNA sources. *Proc. natn. Acad. Sci. U.S.A.* **86**, 6686–6690.

O'Brien, S.J. & Graves, J.A.M. 1991 Report of the committee on comparative gene mapping. *Cytogenet. Cell Genet.* **58**, 1124–1151.

Orbach, M.J., Vollrath, D., Davis, R.W. & Yanofsky, C. 1988 An electrophoretic karytype of Neurospora crassa. *Molec. Cell Biol.* **8**, 1469–1473.

Orita, M., Suzuki, Y., Sekiya, T. & Hayashi, K. 1989 Rapid and sensitive detection of point mutations and DNA polymorphisms using the polymerase chain reaction. *Genomics* **5**, 874–879.

Ott, J. 1974 Estimation of the recombination fraction in human pedigrees: efficient computation of the likelihood for human linkage studies. *Am. J. hum. Genet.* **26**, 588–597.

Paterson, A.H., Damon, S., Hewitt, J.D., Zamir, D., Rabinowtch, H.D., Lincoln, S.E., Lander, E.S. & Tanksley, S.D. 1991 Mendelian factors underlying quantitative traits in tomato: comparison across species, generations and environments. *Genetics* **127**, 181–197.

Paterson, A.H., Lander, E.S., Hewitt, J.D., Peterson, S., Lincoln, S.E. & Tanksley, S.D. 1988 Resolution of quantitative traits into Mendelian factors, using a complete linkage map of restriction fragment length polymorphisms. *Nature, Lond.* **335**, 721–726.

Prochazka, M., Serreze, D.V., Frankel, W.M. & Leiter, E.H. 1991 NOR/Lt Mice MHC-matched diabetes-resistant control strain for NOD mice. *Diabetes* **41**, 98–106.

Riley, J., Butler, R., Ogilvie, D., Finniear, R., Jenner, D., Powell, S., Anand, R. *et al.* 1990 A novel rapid method for the isolation of terminal sequences from artificial chromosome (YAC) clones. *Nucl. Acids Res.* **18**, 2887–2890.

Rommens, J.M., Iannuzzi, M.C., Kerem, B.-S., Drumm, M.L., Melmer, G., Dean, M. & Rozmahel, R. *et al.* 1989 Identification of the cystic fibrosis gene: chromosome walking and jumping. *Science, Wash.* **254**, 1059–1065.

Royle, N.J., Clarkson, R.E., Wong, Z. & Jeffreys, A.J. 1990 Clustering of hypervariable minisatellites in the proterminal regions of human autosomes. *Genomics* **3**, 352–360.

Ruddle, F.H. 1984 The William Allan Memorial Award Address: Reverse genetics and beyond. *Am. J. hum. Genet.* **36**, 944–953.

Schwartz, D.C. & Cantor, C.R. 1984 Separation of yeast chromosome-sized DNAs by pulsed field gradient gel electrophoresis. *Cell* **37**, 67–75.

Searle, A.G., Peters, J., Lyon, M.F., Hall, J.G., Evans, E.P., Edwards, J.H. & Buckle, V.J. 1989 Chromosome maps of man and mouse IV. *Ann. hum. Genet.* **53**, 89–140.

Sefton, L. & Goodfellow, P.N. 1992 The human gene map. *Curr. Top. Genet. Devl Biol.* (In the press.)

Shows, T.B. & McAlpine, P.J. 1979 The 1979 catalogue of human genes and chromosome assignments. *Cytogenet. Cell Genet.* **25**, 117–127.

Solomon, E. & Bodmer, W.F. 1979 Evolution of a sickle variant gene. *Lancet* **1**, 923.

Somerville, C.R. 1989 Arabidopsis blooms. *Plant Cell* **1**, 1135–1137.

Stallings, R.L., Torney, D.C., Hildebrand, C.E., Longmire, J.L., Deaven, L.L., Jett, J.H., Doggett, N.A. & Moyzis, R.K. 1990 Physical mapping of human chromosomes by repetitive sequence fingerprinting. *Proc. natn. Acad. Sci. U.S.A.* **87**, 6218–6222.

Sulston, J., Du, Z., Thomas, K., Wilson, R., Hillier, L., Staden, R., Halloran, N. *et al.* 1992 The C. elegans genome sequencing project: a beginning. *Nature, Lond.* **356**, 37–41.

Tanksley, S.D., Young, N.D., Paterson, A.H. & Bonierbale, M.W. 1989 RFLP mapping in plant breeding: new tools for an old science. *Biotechnology* **7**, 257–266.

Tassabehji, M., Read, A.P., Newton, V.E., Harris, R., Ballings, R., Gruss, P. & Strachan, T. 1992 Waardenburg's syndrome patients have mutations in the human homologue of the Pax-3 paired box gene. *Nature, Lond.* **355**, 635–636.

Tautz, D. & Renz, M. 1984 Simple sequences are ubiquitous elements of eukaryotic genomes. *Nucl. Acids Res.* **12**, 4127–4138.

Taylor, B.A. 1976 Genetic analysis of susceptability to isonazid-induced seizures in mice. *Genetics* **83**, 373–377.

Thoday, J.M. 1961 Location of polygenes. *Nature, Lond.* **191**, 368–370.

Todd, J.A., Aitman, T.J., Cornall, R.J., Ghosh, S., Hall, J.R.S., Hearne, C.M., Knight, A.M. *et al.* 1991 Genetic analysis of autoimmune type 1 diabetes mellitus in mice. *Nature, Lond.* **351**, 542–547.

Verkerk, A.J.M.H., Pieretti, M., Sutcliffe, J.S., Fu, Y.-H., Kuhl, D.P.A., Pizzuti, A., Reiner, O. *et al.* 1991 Identification of a gene (FMR-1) containing a CGG repeat coincident with a breakpoint cluster region exhibiting length variation in Fragile X syndrome. *Cell* **65**, 905–914.

Wallace, R.B., Shaffer, J., Murphy, R.F., Bonner, J., Hirose, T. & Itakura, K. 1979 Hybridization of synthetic oligonucleotides to ox174 DNA: the effect of single base pair mismatch. *Nucl. Acids. Res.* **6**, 3543–3557.

Waugh, R. & Powell, W. 1992 Using RAPD markers for crop improvement. *Trends Biotechnol.* **10**, 186–191.

Weber, J.L. 1990 Informativeness of human $(dC-dA)_n$–$(dG-dT)_n$ polymorphisms. *Genomics* **7**, 524–530.

Weigel, D., Alvarez, J., Smyth, D.R., Yanofsky, M.F. & Meyerowitz, E.M. 1992 *LEAFY* controls floral meristem identity in arabidopsis. *Cell* **69**, 843–859.

Williams, J.G.K., Kubelik, A.R., Livak, K.J., Rafalski, J.A. & Tingey, S.V. 1990 DNA polymorphisms amplified by arbitrary primers are useful as genetic markers. *Nucl. Acids Res.* **18**, 6531–6535.

Wintero, A.K., Fredholm, M. & Thomsen, P.D. 1992 Variable $(dG-dT)_n-(dC-dA)_n$ sequences in the porcine genome. *Genomics* **12**, 281–288.

2

The regulation of sexual development in plants

H. G. Dickinson

SUMMARY

Plant reproduction comprises an interlocking array of developmental pathways which include the formation of the sexual organs, the generation of germ lines *de novo*, and the operation of the mechanisms which regulate epigenetic imprinting and the system of self-incompatibility found in many angiosperms. Little is known of how these processes are regulated at a molecular level, with the exception of the floral organs which are determined by families of homeotic genes operating in a heterochronic fashion. In dioecious and monoecious plants the expression of these 'floral' genes must be modulated by sex-determination sequences, situated in some circumstances on sex chromosomes. Older, physiological data indicate that sex can be determined by growth regulators, particularly gibberellic acid (GA) and cytokinins, and it is possible that sex-determination genes establish local concentrations of growth regulators at the apex, which in turn influence the expression of the homeotic floral genes. Evidence from anther development indicates genes involved in differentiation of the male and female germ lines to be regulated by defined promoter, enhancer, and silencer regions, but few data are available on the sequences directing the initiation and regulation of meiosis; certainly parallels can be drawn with similar events in microorganisms, and useful complementation strategies may be devised, but significant differences do exist between yeasts and higher plants suggesting that more appropriate parallels should be drawn with multicellular eukaryotes such as nematodes. The loci involved in epigenetic imprinting and self-incompatibility are important because they affect both male and female developmental pathways. Nothing is known of the regulatory sequences which direct the epigenetic imprinting of the sperm and central cell genomes, but information is becoming available on the promoter regions of the *S*(incompatibility)-locus. Interestingly, sequences directing expression in male and female tissues are contained within a single 5′ stretch within the locus, and these promoters also induce expression in different cell types in the anther and pistil depending on the type of self-incompatibility involved. Regulation of reproductive development in plants is apparently not very stringent, for there are examples in both male and female germ lines of reversion to an embryonic condition (apomixis and microspore embryogenesis); whether this reflects the highly dedifferentiated state of these cells or differences in the regulation of somatic and reproductive development remains to be determined.

1. INTRODUCTION

Plants and animals differ conspicuously in their sexual strategies, but there are tantalizing similarities in the way sexual differentiation is initiated and regulated in the two kingdoms. However, while key sequences involved in animal sex determination have become focused on identifiable sex chromosomes, the hermaphrodite nature of many flowering plants has not generated the selective pressure for evolution of sexual inheritance mechanisms. The dioecious angiosperms, however, exhibit a range of chromosomally based sexual inheritance systems to rival animals. Even in plants with clear-cut sex chromosomes, explaining sexual determination in terms of gene expression has not proved easy. Certainly dramatic progress is currently being made in the understanding of genes carried by mammalian sex chromosomes (Goodfellow *et al.*, this symposium), but the difficulties inherent in transforming many of the plants involved and the apparent scatter of many of the key genes throughout the genome has resulted in little useful progress to date, although new cloning strategies are beginning to provide encouraging results (Ciupercescu *et al.* 1990; Veuskens *et al.* 1992).

More is known of the molecular basis of sexual differentiation. After induction of flowering, the floral organs themselves have been shown to result from the combinatorial action of a gene family common to the angiosperms (Coen & Meyerowitz 1991; Bradley *et al.*, this symposium). Genes directing the differentiation of plant male germ-line cells have also been characterized (McCormick 1991), and information concerning their promoter and enhancer regions is accumulating (Twell *et al.* 1991). However, at this level, regulation of sexual and somatic development appears identical and the plethora of studies in reproductive development probably reflects the strong commercial interest in modifying plant reproductive systems (Dickinson 1989), and the fact that the anther represents an excellent experimental paradigm, rather than the presence of features unique to reproductive systems.

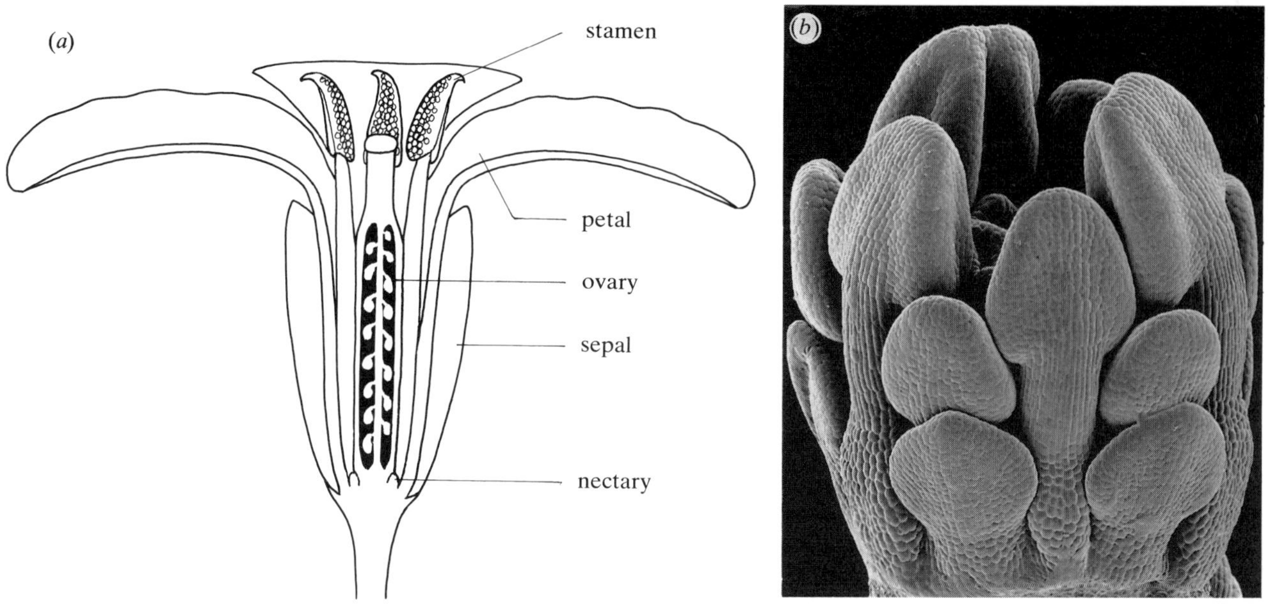

Figure 1. (*a*) Diagrammatic representation of a mature *Brassica* flower showing the principal organs (×8). (*b*) Developing flower of *Silene coeli-rosa*, viewed using low temperature scanning electron microscopy. One whorl of stamens is well advanced whereas another is at an earlier stage of development. The young petals are in register with these smaller stamens (micrograph courtesy of C. Jeffree) (×133).

Interestingly, reproductive development has proved to be surprisingly labile; for example, differentiation of pollen grains may easily be switched to embryogenic development (Guha & Maheshwari 1964; Zaki & Dickinson 1991).

There are, however, two systems that are unique to sexual development and which have different molecular consequences depending upon the sex of the individual: these are epigenetic imprinting and self-incompatibility. Epigenetic, or genomic, imprinting in flowering plants favours sexual reproduction, not by inducing abortion of young embryos in plants attempting parthenogenesis, but by ensuring the failure of the nutritive endosperm tissue (Håkansson 1953). A convincing body of data now exists indicating that the pollen and endosperm genomes receive different epigenetic imprints late in anther and ovary development (Lin 1982; Haig & Westoby 1989).

Ever since the proposition by Lewis (1954) of a tripartite structure for the *S*(incompatibility)-locus, with pollen and stigmatic domains directing the expression of similar products in male cells and the pistil, strenuous attempts have been made to identify such *S*-locus products (Scutt *et al.* 1990). More recent data have dramatically revealed that *S*-locus promoters direct the expression of different products in pollen and stigmas (Sato *et al.* 1991), implying a level of genetic regulation so far unsuspected.

2. SEXUAL DETERMINATION IN HIGHER PLANTS

Sexual expression in flowering plants takes a variety of forms, and this variety is reflected in its genetic control. Most angiosperms carry hermaphrodite flowers, but some 7% are dioecious and 6% monoecious (Irish & Nelson 1989). Selection pressure for the separation of the sexes must have lessened early in the evolution of the angiosperms when the first systems of self-incompatibility evolved (Whitehouse 1950), and thus it is not surprising that sex-chromosome systems have failed to become as prevalent as in animals. Certainly, sex chromosomes are found in dioecious plants, but only in five genera (Parker 1990) of which *Melandrium*, *Rumex*, and *Humulus* have been investigated in some depth. Sex determination in *Melandrium* is truly heterogametic, with XY individuals being male and XX female (Westergaard 1958). The Y chromosome is active, promoting maleness in addition to suppressing female development. In a classical series of experiments involving segregation of parts of the large Y chromosome through triploids, Westergaard was able to demonstrate different domains of the chromosome to regulate enhancement of male development and the suppression of femaleness. Interestingly, as in *Drosophila*, while maleness and femaleness are principally controlled by the Y chromosome, the dosage of X chromosomes or autosomes can modify the maleness expressed (Frankel & Galun 1977). Detailed molecular data are not available either for the Y or X chromosomes of *Melandrium*, but RAPDs have been used to generate a range of Y-specific probes and a number of enrichment strategies have been adopted in attempts to clone Y-specific cDNAs (Veuskens *et al.* 1992).

Both *Rumex* and *Humulus* possess X:autosome balance systems determining sex, but there is considerable variety in each group (Parker & Clarke 1991) with *Rumex hastatulus* displaying elements of both X:autosome and active Y systems. In *Rumex acetosa*, where females are XX and males XY_1Y_2, the X chromosomes are uniform while some considerable variation occurs in Ys, particularly in the centromeric region. Preliminary molecular analysis of the Y chro-

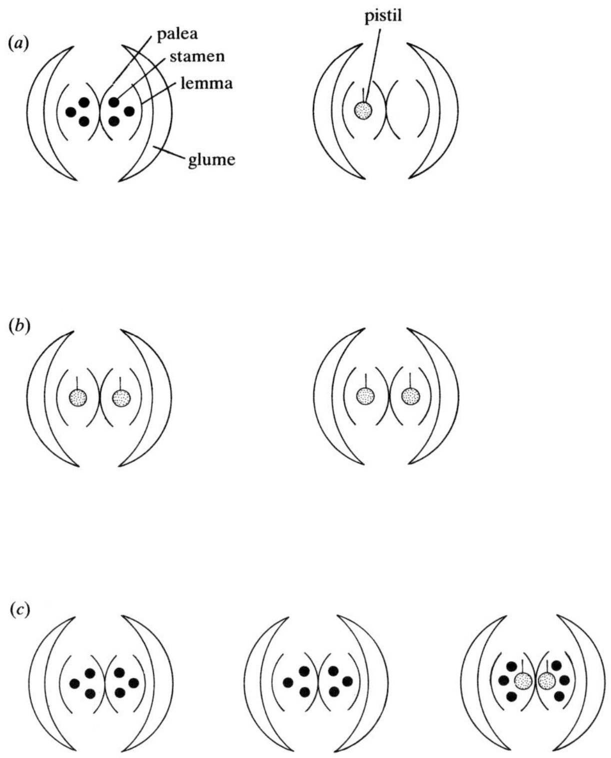

Figure 2. (*a–c*) Diagrammatic representation of floral structure in *Zea mays*, redrawn from Irish & Nelson (1989). (*a*) Wild-type tassel (male; left) and ear (female; right) flowers showing stamens and pistils. (*b*) Tassel seed mutant (*ts-1*, *ts-2*) where stamens have been replaced by pistils. (*c*) Dwarf mutant (*dw*) where pistils have been partially or fully replaced by stamens.

mosomes of *Rumex acetosa* reveals the presence of highly repetitive heterochromatin domains; however, these sequences may also be found in the X chromosomes and scattered throughout the autosomes (Parker & Clarke 1991). Further, the frequency of these sequences in the Y chromosomes does not exceed that found either in the autosomes or X chromosomes, suggesting that the high frequency of heterochromatic sequences seen in the sex chromosomes of mammals (Jones & Singh 1981) is not paralleled in plants. Sex determination in *Rumex acetosa* is via a classical X:autosome balance system, with ratios between X chromosomes and autosome sets higher than 1.0 forming females while ratios lower than 0.5 induce maleness. Intermediate ratios result in 'intersex' plants, as is the case with *Drosophila*. Recent data suggest that the Y chromosome in *Rumex acetosa* is not required for male development *per se*, but affects sporophytic development in such a way as to influence the latter stages of pollen mother-cell differentiation (Parker & Clarke 1991).

Humulus, like *Rumex*, features two types of sex determination system. In *Humulus japonicus* sex is determined by an X:autosome balance system, while *Humulus lupulus* (wild hops), and most cultivated relatives, have a complex, multiple sex-chromosome system (Parker & Clarke 1991). In hops, like *Rumex*, the Y chromosome is not necessary for the development of male flowers, but is required for the maturation of pollen.

Whether determined by genes carried on sex chromosomes or autosomes, sexual determination is achieved by the suppression of development of one set of sex organs. Evidence from work on *Zea* (see Heslop-Harrison (1972), for review) and *Mercurialis* (Durand & Durand 1984) also indicates that growth regulators play a central role in this suppression. In *Zea*, female development is inhibited in tassels and male development in ears. Examination of the earliest stages in floral development indicates the initiation of both sets of sex organs in tassels and ears, but also some other characteristics of the mature structure, for example the branch primordia in tassels (Cheng *et al.* 1983). Two mutants (see figure 2), *dwarf* (*Dw*/*dw*) and *tassel seed* (*Ts*/*ts*), have proved useful in unravelling sex determination in *Zea*. *Dw*/*dw* mutants form normal tassels, but develop male organs in the ears. Six unlinked loci – some dominant and some recessive – have been demonstrated to regulate this mutation; interestingly, at least one recessive can be rescued by the application of exogenous GA (Phinney 1961). The involvement of GA in sex determination in *Zea* is supported by the observation that it will feminize normal tassels (Hansen *et al.* 1976) and that the ear shoots contain some 100 × the somatic levels of GA_1. Further, low light and short days – known to increase GA levels – tend to feminize plants (Heslop-Harrison 1972), and many of the recessive *dw* mutations have now been identified as regulating steps in GA synthesis (Phinney 1961).

In *Ts*/*ts* mutants, pistils develop in the normally staminate tassels. Again there are at least five *Ts*/*ts* mutants, regulated by unlinked loci (Coe *et al.* 1989). While *ts-1* and *ts-2* produce clear phenotypes, other mutations generate complex morphology featuring male and female organs. Significantly the product of at least one of the *tassel seed* alleles is non-diffusible (Johri & Coe 1983), indicating it not to be a growth regulator. The analysis of the few *dwarf*/*tassel seed* double mutants that have been studied suggests the two loci are unlinked. There is also circumstantial evidence that once the sex of a particular flower has been determined, the remainder of development is comparatively stable. Cultured tassels and ears will develop satisfactorily in the absence of GA, although the early stages of female and the later stages of male development require cytokinins (Polowick & Greyson 1985; Bommineri & Greyson 1987). Irish & Nelson (1989) propose that, among other signals, local GA levels are perceived by 'pivotal sex determination genes' that activate appropriate sexual differentiation pathways.

Study of the dioecious *Mercurialis* confirms the central role of growth regulators in the determination of sex. Louis & Durand (1978) have reported three genes to be involved in regulating sexuality. These promote maleness to a greater or lesser extent and were identified by their ability to suppress the feminizing influence of exogenous cytokinins. While earlier experiments had already shown (Durand 1967) that males were feminized by cytokinins, females could only be transformed to males by *in vitro* culture in the presence of auxins (Champault 1973). However, the accumulated growth regulator data suggest that the genes identified by Durant maintain endogenous growth regulator levels, rather than determine sexual expression *per se*. More recently Hamdi *et al.* (1989) have investigated gene expression in male and female flowers of *Mercurialis* using hybridization kinetics and *in vitro* protein synthesis. While their study also involved male sterile and restored lines, specific poly $(A)^+$ RNAs were identified in male flowers, and a cDNA probe was used to demonstrate that this transcript was downregulated in the presence of feminizing hormones (cytokinins). These authors regard these findings as supporting the existence of 'sexual hormones' controlled by regulator genes, which activate the appropriate developmental pathways in bipotential meristems.

The current position with regard to sex determination in plants thus remains highly unsatisfactory. It is not surprising that the adoption of dioecy by groups of plants has led to the evolution of sex chromosomes, but their study, and the search for sequences regulating sex determination in other plants, has resulted in little useful data. Certainly growth regulators appear to be involved in the formation of the sexual organs themselves, but identity of the 'pivotal' genes which activate growth regulator controlling sequences such as *dw* remains unknown. More particularly, the genetic steps are missing which link the gross effects of GA, auxins, and cytokinins to the heterochronic and combinatorial action of the homeotic genes – discussed further below – known to control floral morphogenesis in angiosperms (Coen & Meyerowitz 1991; Bradley *et al.*, this symposium).

3. FLORAL INDUCTION AND INITIATION

The switch to reproductive development in flowering plants is not well understood. Floral initiation normally commences in response to single or multiple environmental cues, and is first detected as cell-cycle changes at the apex (Bernier 1992). Although the switch from vegetative to reproductive development has tacitly been assumed to be regulated by a number of genes, recent studies of natural and transposon-induced mutants suggest otherwise. For example, mutation of the *floricaula* locus in *Antirrhinum* results in indeterminate shoot formation in the sites normally occupied by determinate flowers (Bradley *et al.*, this symposium). Data are not yet available concerning the cytological changes, if any, that take place at the apex of *floricaula* mutants.

Studies of floral mutants have also enabled identification of the genes regulating the development of the induced apex into the complete hermaphrodite flower. This work, which has been carried out simultaneously in *Arabidopsis* and *Antirrhinum* (Coen & Meyerowitz 1991), strongly indicates that – within these groups – floral development is directed by similar sets of genes, and whereas these are expressed sequentially during the generation of the floral organs, they operate in a combinatorial manner in certain phases of development. The formation of individual floral organs is directed by three classes of homeotic genes, encoding functions termed *a*, *b*, and *c*. The *a* function alone results in the formation of sepals; *a* and *b* combined, the production of petals; *b* and *c*, stamens; and *c* alone, carpels. Interestingly, function *c* – revealed, for example, by mutations at the *agamous* locus in *Arabidopsis* – also regulates floral determinacy. While *Arabidopsis* is radially symmetrical *Antirrhinum* is zygomorphic, indicating the presence of genes which modulate the *a*, *b* and *c* functions. Mutations at the *cycloidia* locus in *Antirrhinum* result in the formation of radially symmetrical flowers and the products of this locus are considered to interact with the organ-generating functions in particular domains of the developing apex (Carpenter & Coen 1990).

Although all of these genes have yet to be fully characterized, some (e.g. *deficiens* from *Antirrhinum*) have been shown to encode transcription factors with homology to sequences in yeast and mammals (Coen & Meyerowitz 1991). Of particular interest is the role

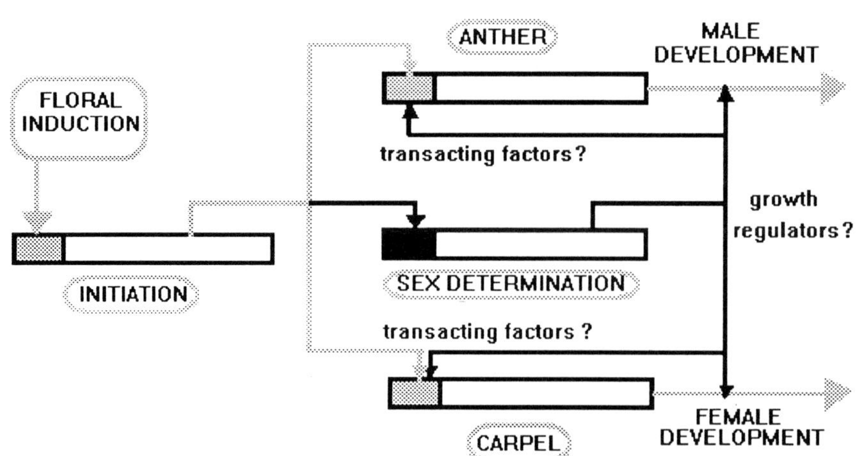

Figure 3. Regulation of sex determination in higher plants. Sex determination genes, carried on sex chromosomes or scattered throughout the genome, are activated following floral initiation. The products of these genes modulate floral organogenesis either directly by gene regulation, or indirectly through altering local concentrations of growth regulators.

played by these genes in monoecy and dioecy. The formation of single-sex flowers has always been regarded as involving the suppression of either stamens or carpels; conditions which can be mimicked in the mutants described above. Further, both hormonal treatments and environmental stress can induce changes in sex expression in many plants. Partial answers to these questions about sex determination may be provided by *in situ* hybridization, which enables the expression of the gene products of many of these sequences to be followed at a cellular level (Jones *et al.* 1990), but it remains unclear whether 'sex-determination' sequences (on sex chromosomes or otherwise) modulate the expression of the organ-forming homeotic genes such as *agamous* and *sepaloidia* directly by *trans*-acting factors, or through local changes in the levels of growth regulators (see figure 3). For this reason, the nature of the sequences carried on the sex determining domains of active Y chromosomes, such as those of *Melandrium*, remains of considerable interest.

4. MALE AND FEMALE GERM-LINE DEVELOPMENT

Development of the male germ line and its investing tapetal tissue in the anther has been the focus of much recent molecular research. Although development of the embryo sac and its nutritive nucellus is of equal significance, a combination of the small numbers of cells involved, their inaccessibility, and a lack of commercial interest has resulted in female gametogenesis remaining largely unstudied. However, the use of *in situ* and immunolocalization methods is permitting the investigation of molecular events in the differentiating ovule (Dow & Mascarenhas 1991). A striking difference between the germ lines of animals and plants is that the latter are formed *de novo*. Gametes in higher animals are normally derived from a distinct cell line, determined in the embryo and giving rise only to cells involved in reproduction. In plants, following floral induction, vegetative meristems alter in their organization, forming flowers. Within the flowers, gametes develop in the ovaries and anthers.

Conservative estimates indicate the presence of 10 000 anther-specific genes (Kultnow *et al.* 1990) and thus it is not surprising that differential screening and the screening of subtractive libraries have resulted in the characterization of anther-specific cDNA, ranging in time of expression from meiosis to late in pollen development (see McCormick (1991) for review). A number of genes have also been cloned from mature and germinating pollen (Mascarenhas 1992). Interestingly, most sequences have been cloned from pollen mitosis I onwards, presumably reflecting the low levels of poly $(A)^+$ RNA in meiotic cells (Porter *et al.* 1984). Further, the intense synthetic activity of the tapetum is reflected in the large number of tapetal-specific cDNAs now reported (McCormick 1991), although there has been commercial pressure to obtain the promoters for sequences of this type for use in engineered systems of male sterility (Mariani *et al.* 1990). More recent data suggest that a small but significant number of anther-specific sequences, of which the *Bcp1* gene from *Brassica* is an example (Therakulpisut *et al.* 1991), are expressed in both the sporogenous cells and the tapetum, generally at different developmental stages. Not surprisingly, a large proportion of the sequences expressed in the anther, such as *LAT52* from *Lycopersicum* (McCormick 1991), are also transcribed in other organs and tissues. Despite this plethora of anther-specific genes, those directing the initiation and progression of meiosis remain elusive. Apart from the *pLEC* sequences cloned by Bouchard (1990), which appear to belong to a heat-shock family, no meiosis-specific genes have been identified, although their homologues in yeast are well characterized (e.g. *HOP1*, Hollingsworth *et al.* 1990). It is thus likely that the genes involved in meiotic initiation and recombination will be cloned by complementation, or through their homology at the DNA or protein level with sequences from organisms such as *Saccharomyces* and *Caenorhabditis*.

The polypeptides encoded by anther-specific genes have been predicted from DNA sequence data and range from ascorbate oxidase, via pectic lyases and polygalacturonases, to lipid transfer proteins and Kunitz trypsin inhibitors (see McCormick (1991) for review). Kopczak *et al.* (1992) have recently reported specific forms of β tubulin to be synthesized in the pollen, and it is possible that there are pollen-specific forms of other cytoskeletal components including actin and intermediate filaments. A surprising number of pollen-specific cDNAs, while encoding putative proteins rich in specific amino acids (particularly glycine, proline and cysteine), have no homologies with sequences on current data bases. The pectin-degrading enzymes and polygalacturonases presumably play a role in the development of pollen on the stigma surface and in the style, but otherwise few specific functions have been assigned to pollen- or tapetum-specific genes.

The most comprehensive analysis of the promoter regions of pollen-specific genes has been carried out by Twell *et al.* (1991) on *LAT52* and *LAT59* from *Lycopersicum*. These genes are coordinately expressed late in anther development, both in the pollen and sporophytic tissue. The expression in transgenic plants of constructs containing deleted promoter sequences linked to the *GUS* reporter gene demonstrated only small stretches of promoter sequence (less than 200 b.p.) to be necessary for expression in pollen and specific sporophytic cells. Twell *et al.* (1991) also used a transient assay system to identify a number of upstream *cis*-acting regulatory regions of these, and other, promoters. Interestingly, one of these regions enhanced the expression of the *CaMV35S* promoter in pollen. Analysis of these *cis* elements from different promoters revealed shared sequences, suggesting that similar *trans*-acting factors may activate a number of anther-specific genes. Most recently, Twell (1992) has used the promoter region of the *LAT52* gene from *Lycopersicum* to investigate gene regulation in the developing microgametophyte. Again using a *LAT52-GUS* construct, but on this occasion fusing a nuclear location signal to the *GUS* gene, he demonstrated the promoter to direct expression of the reporter gene in the vegetative cell in transgenic *Nicotiana*. GUS was detected neither in the nucleus nor in the cytoplasm of the generative cell. This dramatic difference in gene expression in adjacent cells, so soon after their formation from a common parent, underlines the significance of pollen mitosis I in male germ-line development (Dickinson 1992*a*). As long ago as 1964 Guha & Maheshwari demonstrated that disruption of this division could 'reset' development and result in the production of embryos. More recent data (Pechan & Keller 1988; Zaki & Dickinson 1990, 1991) suggest that the asymmetry of this division is crucial to the correct development of both vegetative and generative cells. Treatment with colchicine immediately prior to pollen mitosis I, which induces division symmetry, results in a higher rate of embryogenesis (Zaki & Dickinson 1991), suggesting that simple factors such as nuclear:cytoplasmic volume ratios may determine the subsequent fates of these cells. The molecular basis of this developmental switch has yet to be determined, but Pechan *et al.* (1991) have demonstrated heat-shock proteins to be synthesized in microspores after inductive treatments. These polypeptides, which can interact with the cytoskeleton, may act directly to disrupt the symmetry of division, or operate via gene expression to switch development from a gametophytic to a sporophytic pathway.

Developmental flexibility, of which microspore embryogenesis is an example, is a characteristic of reproductive development in higher plants. It is particularly evident in apomictic development in females, where embryos can arise spontaneously from either the nucellar tissue or from one of the cells of the embryo sac. Sometimes apomixis is stimulated by pollination, but in most instances it occurs spontaneously. Many angiospermous groups contain facultative or constitutive apomicts and a surprisingly high proportion of higher plant seed is produced by this process. The genetic control of apomixis is poorly understood, but t-DNA mutation of *Arabidopsis* has resulted in apomictic plants (E. Herbele-Bors, personal communication) and thus the way is open for the identification and characterization of genes involved in this process.

5. EPIGENETIC IMPRINTING DURING SEXUAL DIFFERENTIATION

The presence of a mechanism promoting double fertilization in angiosperms by haploid gametic nuclei was first indicated in experiments involving reciprocal pollinations between individuals of different ploidy. Håkansson (1953), working on *Brassica*, demonstrated that pollination with diploid gametes resulted in the inhibition of endosperm development and lack of seed set. Interestingly, matings between individuals of increased ploidy were sometimes successful, especially when a 'normal' gamete:endosperm (polar) nuclear ratio of 1:2 was maintained. Study of the expression of the *R*-locus of *Zea* also produced some puzzling results, for different phenotypes could be produced by the same genotypic combination of dominant and recessive alleles, depending upon whether particular alleles were donated by the male or female parent. As this could be interpreted as a gene dosage effect, Kermicle (1970) used the B chromosome reciprocal exchange system in *Zea* to investigate whether this apparent asymmetry in gene expression resulted from a wrong balance of alleles being present in the endosperm, or from the fact that the maternal:paternal nuclear contributions were incorrect. The results proved conclusively that the *R*-locus phenotype was regulated by the parental origin of the alleles involved. The long arm of chromosome 10 in *Zea* carries a number of genes involved in endosperm development and Lin (1982), again using the B chromosome translocation system, synthesized six interchanges with reduced endosperm development. Interestingly, endosperms of all genetic constitutions, but with appropriate dosages of chromosome 10 donated maternally rather than from both parents, proved to be defective. Thus a proportion of the endosperm-specific sequences on chromosome 10 must be donated paternally for effective development of the tissue.

Although this sex-specific imprinting of the genome undoubtedly favours sexual reproduction, and may even be significant in the apomictic strategies adopted by plants (Haig & Westoby 1989), there is no information as to how it is achieved. The early view that epigenetic imprinting in animals involved differential methylation of key sequences is currently under serious reconsideration (Surani *et al.*, this symposium). Certainly many sequences – including transgenes – demonstrated to be imprinted are heavily methylated, but on the other hand not all methylated sequences are imprinted, at least as far as determining the fate of the young embryo. Methylation of plant DNA is well documented, and changes in the methylation status of particular sequences have been reported to result in alterations in their expression (Heslop-Harrison 1990). Indeed, selective demethylation of specific endosperm genes during reproductive development may well represent the mechanism by which these genes are activated (Bianchi & Viotti 1988). However, as is the case with animals, conclusive data have yet to be obtained showing that differential imprinting during plant sexual development is achieved solely through cytosine methylation. Most recently, families of genes have been identified of which the products determine the supramolecular structure within the chromosome, and thus, indirectly, can regulate gene expression. Perhaps the best characterized is *polycomb* in *Drosophila* (Jürgens 1985), the gene product of which is responsible for the organization of heterochromatin domains, in which gene expression is suppressed. A *polycomb* homologue has been identified in plants (Singh *et al.* 1991), and it remains a distinct possibility that a combination of alterations in chromatin structure, induced by the action of *polycomb*-like genes, and selective methylation results in differential epigenetic imprints on the sperm and central cell genomes.

The juncture at which the male and female genomes are imprinted is not known. The sperms in the pollen (or pollen tube) are derived from the generative cell which is formed, with the vegetative cell, by the asymmetrical division of the microspore. Because the microspore, the vegetative cell, and sometimes even the generative cell may be switched directly into an embryogenic pathway (Dickinson 1992*b*), the epigenetic imprint – if at all comprehensive – is most likely applied during the maturation of the highly condensed sperm nuclei. Because the character and fate of the binucleate central cell differs so markedly from that of the other cells of the embryo sac, any epigenetic modifications must take place during the final differentiation of the polar nuclei. The differential epigenetic imprinting involved in higher plant reproductive development thus represents part of the general sequence of sexual differentiation, presumably regulated by promoters – so far undiscovered – directing expression in the sperm cells of the pollen and the central cell of the embryo sac.

6. THE REGULATION OF GENE EXPRESSION IN SELF-INCOMPATIBILITY

The *S*(incompatibility)-locus is of considerable interest because, although specific to reproductive development, it is differentially expressed in the cells of the pollen and pistil. In the majority of angiosperms possessing self-incompatibility (SI) the presence of identical alleles at a single *S*-locus results in the identification and rejection of self-pollen (Dickinson 1990) (see figure 4). There are considerable variations on this theme; the Graminae possess two loci and *Ranunculus* spp. three. Further, heteromorphic SI mechanisms are regulated by complex, single *S*-loci which also encode 'physiological' self-rejection systems.

SI in members of the Cruciferae and Asteraceae is also complicated by the sporophytic inheritance of pollen compatibility (Bateman 1955), where the alleles of the pollen parent, rather than of the haploid microgametophyte itself, determine the fate of the grain on the stigma surface. Although the site of the *S*-allele products in the pollen is so far unknown, their sporophytic origin suggests that they are synthesized prior to meiosis and survive the entire gametophytic phase, or that they are applied to the pollen late in development, perhaps as part of the tapetally derived pollen grain coating (Dickinson & Lewis 1973).

The elegant dimer hypothesis of Lewis (1965) predicts that identical *S*-locus products are formed in the pollen and pistil and that, on pollination, a dimer is formed which inhibits further development. Other domains of the locus are held to direct the expression of messages encoding these products in male and female tissues. The use of then-new cloning strategies by Anderson *et al.* (1986) and Nasrallah *et al.* (1985) resulted in the identification and characterization of *S*-locus sequences encoding glycoproteins (*S*-locus glycoprotein genes: *SLG*s) expressed in the pistils of *Nicotiana* and *Brassica* respectively. More recent work on the *S*-locus of *Brassica oleracea* has revealed it also to contain an *S*-receptor kinase (*SRK*) gene (Stein *et al.* 1991) analogous to a transmembrane serine–threonine kinase gene found in *Zea* (Walker & Zhang 1990). In addition, sequences with strong homologies to the *SLG*s are also present elsewhere in the genome, but

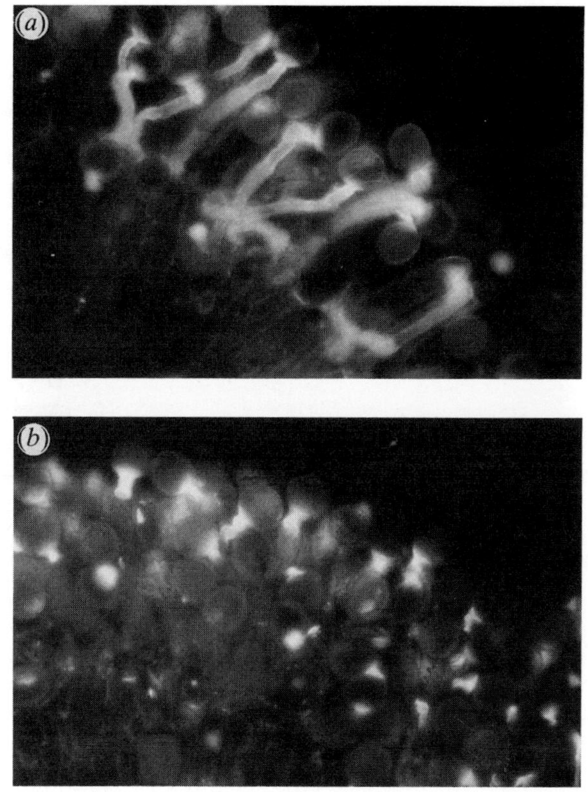

Figure 4. (*a,b*) Self-incompatibility in *Brassica oleracea*. (*a*) Compatible cross-pollination stained with aniline blue and viewed in the fluorescence microscope. Fluorescence of the callose in the pollen tube walls shows the pollen to have germinated and the tubes to be entering the pistillar tissue. (*b*) As (*a*), but following a self-pollination. Pollen grains have either failed to germinate, or produced very truncated tubes. (× 660)

unlinked to the *S*-locus (Lalonde *et al.* 1989; Trick & Flavell 1989). While the *SRK* is expressed in both male and female cells, the male determinant of SI in *Brassica* has yet to be unequivocally identified, although peptides derived from the pollen grain coating have been demonstrated to interact specifically with the *SLG* product (Doughty *et al.* 1993*a,b*).

The study of promoters within the *S*-locus has proved particularly revealing. Transgenic expression, in *Brassica oleracea*, of constructs containing regions upstream of *SLG* coupled with various reporter genes indicates these sequences to direct expression solely in the superficial papillar cells of the stigma, and to a far lesser extent in the vestigial transmitting tissue of this plant (Sato *et al.* 1991). More dramatically, these promoters also direct expression of the reporter sequences in the anther, both in the tapetal cells and less strongly in the microspores (Sato *et al.* 1991). This observation is significant not only because expression occurs in the tapetum, which may be the source of the *S*-allele products carried by the pollen, but also because – contrary to expectation – a single stretch of 5′ upstream sequences seems to be involved in regulating expression in male and female cells. While it is possible that these promoters respond to different

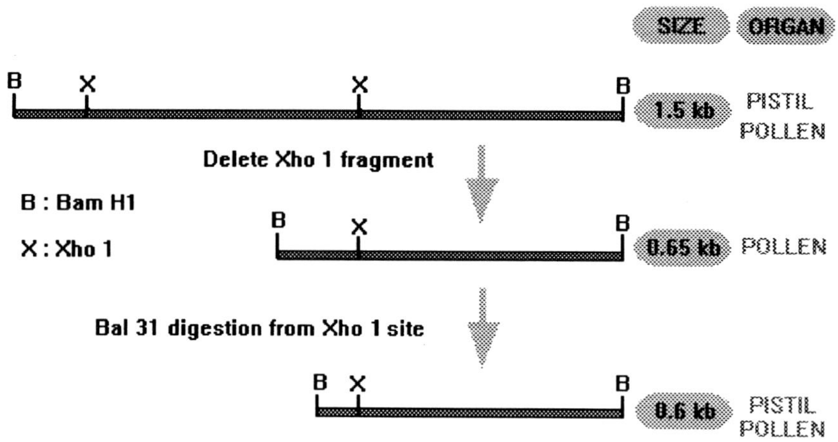

Figure 5. Expression of the *GUS* reporter gene in *Nicotiana*, driven by modified *Brassica SLR* promoters. The full 1.5 kb 5′ region (top) directs expression in the pollen and pistil at appropriate times. Deletions to generate a 0.65 kb promoter (middle) result in pollen expression only. Further truncation of the promoter to 0.6 kb (bottom) resulted in expression in pollen and pistils, but with some loss of temporal regulation (data from Hackett *et al.* 1992).

trans-acting factors in anthers and stigmas, and indeed other sequences are involved in the expression of the male and female *S*-locus products, the effectiveness of reporter constructs in cells of both sexes suggests that SI has evolved as a truly pan-sexual system. Thus although SI may well have evolved from a pathogen defence system located in the pistil (Dickinson *et al.* 1992), gene modification and duplication within the locus, combined with the development of *trans*-acting factors capable of activating it within the pollen, have resulted in the generation of male and female determinants capable of rejecting self-pollen.

Transgenic expression of reporter constructs containing *Brassica SLG* promoters have provided further unexpected data, for in *Nicotiana* (where the inheritance of SI is gametophytic) the reporter gene is expressed in the transmitting tissue of the style, rather than in the stigma, and in the developing pollen instead of the tapetal cells (Thorsness *et al.* 1991). Significantly, in gametophytically regulated SI, the *S*-locus products are expressed in the transmitting tissue, and presumably in the pollen. Thus the *trans*-acting factors responsible for the activation of what appears to be a common SI promoter are distributed in a cell-specific fashion so as to ensure the appropriate expression of the *S*-locus.

Brassica SLG promoters are under study in a number of laboratories, but preliminary results suggest that like many reproductive cell-specific promoters (Twell *et al.* 1991), these sequences contain a

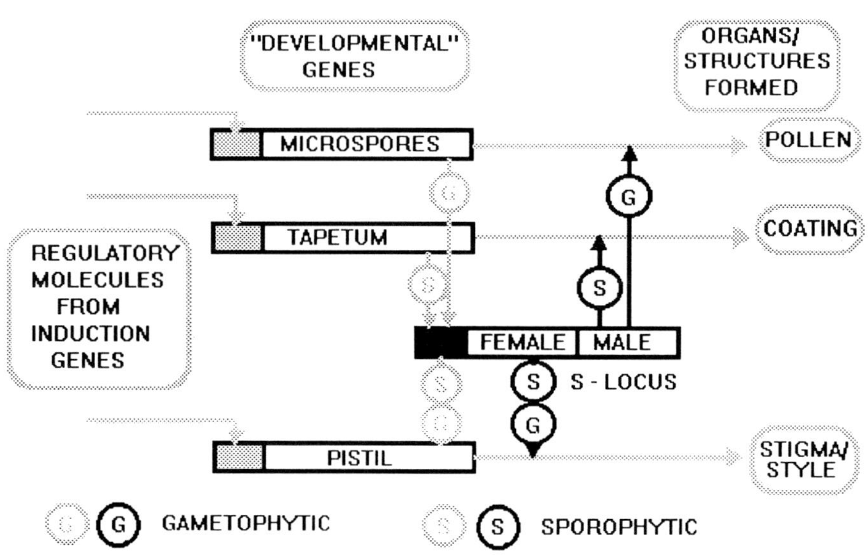

Figure 6. *S*(incompatibility)-locus expression in angiosperms. The locus is differently regulated in sporophytically and gametophytically controlled SI. In sporophytic SI the male *S*-product appears to be synthesized in the tapetum, while in gametophytic SI it is formed in the pollen. Presumably the locus is activated by factors in these different cell types.

mosaic of enhancers and suppressors ranging from a few to tens of bases in length (Hackett *et al.* 1992). No definitive consensus sequence has yet been identified which will direct the expression of the *S*-locus under all circumstances. However, promoter deletion experiments involving the components of the 1.5 kb upstream region of the *SLR-1* (a sequence coexpressed with *SLG*) *Brassica* gene driving a *GUS* reporter construct in transgenic *Nicotiana* (see figure 5), indicate the presence both of sequences which direct expression in male and female tissues, and of silencing elements (Hackett *et al.* 1992). The fact that the same sequence will direct expression in different tissues in plants with differing systems of SI is interesting (see figure 6), and points either to these promoters having a previously unsuspected complexity, or to differences in the distribution of similar *trans*-acting factors.

The author would like to thank Ann Rogers and John Baker for help in preparation of the manuscript. Advice, unpublished data, and illustrative material from very many sources and individuals are also gratefully acknowledged.

REFERENCES

Anderson, M.A., Cornish, E.C., Mau, S.L., Williams, E.G., Hoggart, R., Atkinson, A., Bonig, I., Grego, B., Simpson, R., Roche, P.J., Haley, J.D., Niall, H.D., Tregear, G.W., Coghlan, J.P., Crawford, R.J. & Clarke, A.E. 1986 Cloning of cDNA for a stylar glycoprotein associated with expression of self-incompatibility in *Nicotiana alata*. *Nature, Lond.* **321**, 38–44.

Bateman, A.J. 1955 Self-incompatibility systems in angiosperms. III. Cruciferae. *Heredity, Lond.* **9**, 52–68.

Bernier, G. 1992 The control of flower formation: exogenous and endogenous. In *Reproductive biology and plant breeding* (ed. Y. Dattée, C. Dumas & A. Gallais), pp. 29–39. Berlin: Springer-Verlag.

Bianchi, M.W. & Viotti, A. 1988 DNA methylation and tissue-specific transcription of the storage protein genes of maize. *Plant Molec. Biol.* **11**, 203–214.

Bommineri, V.R. & Greyson, R.I. 1987 *In vitro* culture of ear shoots of *Zea mays* and the effect of kinetin on sex expression. *Am. J. Bot.* **74**, 883–890.

Bouchard, R.A. 1990 Characterisation of expressed inactive transcript clones of *Lilium*; meiosis specific expression, relatedness and affinity to small heat shock protein genes. *Genome* **33**, 68–79.

Carpenter, R. & Coen, E.S. 1990 Floral homeotic mutations produced by transposon mutagenesis in *Antirrhinum majus*. *Genes Dev.* **4**, 1483–1493.

Champault, A. 1973 Effet de quelques régulateurs de la croissance sur des noeuds isolés de *Mercurialis annua* L. (2n = 16) cultivés *in vitro*. *Bull. Soc. Bot. Fr.* **120**, 87–100.

Cheng, P.C., Greyson, R.I. & Walden, D.B. 1983 Organ initiation and the development of unisexual flowers in the tassel and ear of *Zea mays*. *Am. J. Bot.* **70**, 450–462.

Ciupercescu, D.D., Veuskens, J., Mouras, A., Ye, D., Briquet, M. & Negrutiu, I. 1990 Karyotyping *Melandrium album*, a dioecious plant with heteromorphic sex chromosomes. *Genome* **33**, 556–562.

Coe, E.H., Neutter, M.G. & Hoisington, D.A. 1989 The genetics of corn. In *Corn and corn improvement*, 3rd edn. (ed. G. F. Sprague & J. Dudley), pp. 81–258. Madison, WI: American Society of Agronomy.

Coen, E.S. & Meyerowitz, E.M. 1991 The war of the whorls: genetic interactions controlling flower development. *Nature, Lond.* **353**, 31–37.

Dickinson, H.G. (ed.) 1989 *Core biology of plant breeding*. (28 pages.) Review prepared for the Commission of the European Communities.

Dickinson, H.G. 1990 Self-incompatibility in flowering plants. *Bioessays* **12**, 155–161.

Dickinson, H.G. 1992*a* Gene expression during microsporogenesis. In *Plant sexual reproduction* (ed. M. Cresti & A. Tiezzi), pp. 1–17. Springer Verlag.

Dickinson, H.G. 1992*b* Microspore derived embryogenesis. In *Plant sexual reproduction* (ed. M. Cresti & A. Tiezzi), pp. 17–31. Springer Verlag.

Dickinson, H.G. & Lewis, D. 1973 The formation of the tryphine coating the pollen grains of *Raphanus* and its properties relating to the self-incompatibility system. *Proc. R. Soc. Lond.* B **184**, 149–165.

Dickinson, H.G., Crabbe, M.J.C. & Gaude, T. 1992 Sporophytic self-incompatibility systems: *S*-gene products. *Int. Rev. Cytol.* **140**, 525–565.

Doughty, J., McCubbin, A., Hedderson, F., Elleman, C.J. & Dickinson, H.G. 1993*a* The role of the pollen grain coating in pollination and self-incompatibility in *Brassica oleracea*. (In the press.)

Doughty, J., Hedderson, F., McCubbin, A. & Dickinson, H.G. 1993*b* Interactions between a coating-borne peptide of the *Brassica* pollen grain and *S*(incompatibility)-locus linked stigmatic glycoproteins. *Proc. natn. Acad. Sci. U.S.A.* (In the press.)

Dow, D.A. & Mascarenhas, J.P. 1991 Synthesis and accumulation of ribosomes in individual cells of the female gametophyte of maize during its development. *Sex. Pl. Reprod.* **4**, 250–254.

Durand, B. 1967 L'Expression du sexe chez les *Mercuriales annuelles*. *Bull. Soc. Fr. Physiol. Veg.* **13**, 195–202.

Durand, R. & Durand, B. 1984 Sexual differentiation in higher plants. *Physiologia Pl.* **60**, 267–274.

Frankel, R. & Galun, E. 1977 *Pollination mechanisms, reproduction, and plant breeding*. Berlin: Springer-Verlag.

Guha, S. & Maheshwari, S.C. 1964 *In vitro* production of embryos from anthers of *Datura*. *Nature, Lond.* **204**, 497.

Hackett, R.M., Lawrence, M.J. & Franklin, F.C.H. 1992 A *Brassica* *S*-locus related gene promoter directs expression in both pollen and pistil of tobacco. *Plant J.* **2**, 613–617.

Haig, D. & Westoby, M. 1989 Parent specific gene expression and the triploid endosperm. *Am. Nat.* **134**, 147–155.

Håkansson, A. 1953 Endosperm formation after 2*x*, 4*x* crosses in certain cereals, especially in *Hordeum vulgare*. *Hereditas* **39**, 57–64.

Hamdi, S., Yu, L.X., Cabré, E. & Delaigue, M. 1989 Gene expression in *Mercurialis annua* flowers: In vitro translation and sex genotype specificity. Male-specific cDNA cloning and hormonal dependence of a corresponding specific RNA. *Molec. Gen. Genet.* **219**, 168–176.

Hansen, D.J., Bellman, S.K. & Sacher, R.M. 1976 Gibberellic acid-controlled sex expression in corn tassels. *Crop Sci.* **16**, 371–374.

Heslop-Harrison, J. 1972 Sexuality of angiosperms. In *Plant physiology*, vol. VIC (*Physiology of development: from seeds to sexuality*) (ed. F. C. Steward), pp. 133–289. New York: Academic Press.

Heslop-Harrison, J.S. 1990 Gene expression and parental dominance in hybrid plants. *Development* (Suppl. 1990), 21–28.

Hollingsworth, N.M., Goetsch, L. & Byers, B. 1990 The *HOP1* gene encodes a meiosis-specific component of yeast chromosomes. *Cell* **61**, 73–84.

Irish, E.E. & Nelson, T. 1989 Sex determination in monoecious and dioecious plants. *Pl. Cell* **1**, 737–744.

Johri, M.M. & Coe, E.H., Jr 1983 Clonal analysis of corn plant development. I. The development of the tassel and the ear shoot. *Dev. Biol.* **97**, 154–172.

Jones, K.W. & Singh, L. 1981 Conserved sex-associated repeated DNA in vertebrates. In *Genome evolution* (ed. G. Dover & R. Flavell), pp. 135–154. London: Academic Press.

Jones, K., Crossley, S. & Dickinson, H.G. 1990 Investigation of gene expression during plant gametogenesis as revealed by *in situ* hybridisation using non-isotopic probes. In: In situ *hybridisation and the study of development and differentiation* (ed. N. Harris & D. Wilkinson), pp. 189–203. SEB Seminar Series. Cambridge University Press.

Jürgens, G. 1985 A group of genes controlling the spatial expression of the bithorax complex in *Drosophila*. *Nature, Lond.* **316**, 153–155.

Kermicle, J.L. 1970 Dependence of the R-mottled aleurone phenotype in maize on mode of stomal transmission. *Genetics* **66**, 69–85.

Kopczak, S.D., Haas, N.A., Hussey, P.J., Silflow, C.D. & Snustad, D.P. 1992 The small genome of *Arabidopsis* contains at least six expressed α-tubulin genes. *Pl. Cell* **4**, 539–547.

Kultnow, A.M., Truettner, J., Cox, K.H., Wallroth, M. & Goldberg, R.B. 1990 Different temporal and spatial gene expression patterns occur during anther development. *Pl. Cell* **2**, 1201–1224.

Lalonde, B.A., Nasrallah, M.E., Dwyer, K.G., Chen, C.H., Barlow, B. & Nasrallah, J.B. 1989 A highly conserved *Brassica* gene with homology to the S-locus-specific glycoprotein structural gene. *Pl. Cell* **1**, 249–258.

Lewis, D. 1954 Comparative incompatibility in angiosperms and fungi. *Adv. Genet.* **6**, 235–285.

Lewis, D. 1965 A protein dimer hypothesis on incompatibility. In *Genetics today* (ed. S. J. Geerts), pp. 657–663. *Proc. XI Int. Congress Genet. 1963.* Oxford: Pergamon.

Lin, B.-Y. 1982 Association of endosperm reduction with parental imprinting in maize. *Genetics* **100**, 475–486.

Louis, J.P. & Durand, B. 1978 Studies with the dioecious angiosperm *Mercurialis annua* L. (2n = 16); Correlation between genetic and cytoplasmic male sterility, sex segregation and feminising hormones (cytokinins). *Molec. Gen. Genet.* **165**, 309–322.

Mariani, C., de Beuckeleer, M., Truettner, J., Leemans, J. & Goldberg, R.B. 1990 Induction of male sterility in plants by a chimaeric ribonuclease gene. *Nature, Lond.* **347**, 737–741.

Mascarenhas, J.P. 1992 Gametophytic gene expression. In *Reproductive biology and plant breeding* (ed. Y. Dattée, C. Dumas & A. Gallais), pp. 69–77. Berlin: Springer-Verlag.

McCormick, S. 1991 Molecular analysis of male gametogenesis in plants. *Trends Genet.* **7**(9), 298–303.

Nasrallah, J.B., Kao, T.-H., Goldberg, M.L. & Nasrallah, M.E. 1985 A cDNA clone encoding an S-locus specific glycoprotein from *Brassica oleracea*. *Nature, Lond.* **318**, 263–267.

Parker, J.S. 1990 Sex chromosomes and sexual differentiation in flowering plants. *Chrom. Today* **10**, 187–198.

Parker, J.S. & Clarke, M.S. 1991 Dosage sex chromosome systems in plants. *Pl. Sci.* **80**, 79–82.

Pechan, P.M. & Keller, W.A. 1988 Identification of potentially embryogenic microspores in *Brassica napus* L. *Physiologia Pl.* **74**, 377–384.

Pechan, P.M., Bartels, D., Brown, D.C.W. & Schell, J. 1991 Messenger-RNA and protein changes associated with induction of *Brassica* microspore embryogenesis. *Planta* **184**, 161–165.

Phinney, B.O. 1961 Dwarfing genes in *Zea mays* and their relation to the gibberellins. In *Plant growth regulation* (ed. R. J. Klein), pp. 489–501. Iowa State University Press.

Polowick, P.L. & Greyson, R.I. 1985 Microsporogenesis and gametophyte maturation in cultured tassels of *Zea mays* L. *Can. J. Bot.* **63**, 2196–2199.

Porter, E.C., Parry, D., Bird, J. & Dickinson, H.G. 1984 Nucleic acid metabolism in the nucleus and cytoplasm of angiosperm meiocytes. In *Controlling events in meiosis* (ed. C. Evans & H. G. Dickinson), pp. 363–369. Cambridge: Company of Biologists.

Sato, T., Thorsness, M.K., Kandasamy, M.K., Nishio, T., Hirai, M., Nasrallah, J.B. & Nasrallah, M.E. 1991 Activity of an S-locus gene promoter in pistils and anthers of transgenic *Brassica*. *Pl. Cell* **3**, 867–876.

Scutt, C.P., Gates, P.J., Gatehouse, J.A., Boulter, D. & Croy, R.D.D. 1990 A cDNA encoding an S-locus specific glycoprotein from *Brassica oleracea* plants containing the S5 self-incompatibility allele. *Molec. Gen. Genet.* **220**, 409–413.

Singh, A., Ai, Y. & Kao, T.-H. 1991 Characterization of ribonuclease activity of three *S*-allele-associated proteins of *Petunia inflata*. *Pl. Physiol.* **96**, 61–68.

Singh, P.B., Miller, J.R., Pearce, J., Kothary, R., Burton, R.D., Paro, R., James, T.C. & Gaunt, S.J. 1991 A sequence motif in a *Drosophila* heterochromatin protein is conserved in animals and plants. *Nucl. Acids Res.* **19**, 789–794.

Stein, J.C., Howlett, B., Boyes, D.C., Nasrallah, M.E. & Nasrallah, J.B. 1991 Molecular cloning of a putative receptor protein kinase gene encoded at the self-incompatibility locus of *Brassica oleracea*. *Proc. natn. Acad. Sci. U.S.A.* **88**, 8816–8820.

Therakulpisut, P., Xu, H., Singh, M.B., Pettitt, J.M. & Knox, B. 1991 Isolation and developmental expression of Bcp1, an anther-specific cDNA clone in *Brassica campestris*. *Pl. Cell* **3**, 1073–1084.

Thorsness, M.K., Kandasamy, M.K., Nasrallah, J.B. & Nasrallah, M.E. 1991 A *Brassica* S-locus gene promoter targets toxic gene expression and cell death to the pistil and pollen of transgenic *Nicotiana*. *Devl. Biol.* **143**, 173–184.

Trick, M. & Flavell, R.B. 1989 A homozygous S-genotype of *Brassica oleracea* expresses two S-like genes. *Molec. Gen. Genet.* **218**, 212–217.

Twell, D. 1992 Use of a nuclear targeted β-glucuronidase fusion protein to demonstrate cell-specific gene expression in developing pollen. *Plant J.* **2**, 887–892.

Twell, D., Yamaguchi, J., Wing, R.A., Ushiba, J. & McCormick, S. 1991 Promoter analysis of genes that are coordinately expressed during pollen development reveals pollen-specific enhancer sequences and shared regulatory elements. *Genes Dev.* **5**, 496–507.

Veuskens, J., Lacroix, C., Truong, A.T., Hinnisdaels, S., Mouras, A. & Negrutiu, I. 1992 Genetic and molecular enrichment steps as cloning strategy in the dioecious *Melandrium album* (*Silene alba*). In *Reproductive biology and plant breeding* (ed. Y. Dattée, C. Dumas & A. Gallais), pp. 39–49. Berlin: Springer-Verlag.

Walker, J.C. & Zhang, R. 1990 Relationship of a putative receptor protein kinase from maize to the S-locus glycoproteins of *Brassica*. *Nature, Lond.* **345**, 743–746.

Westergaard, M. 1958 The mechanism of sex determination in dioecious flowering plants. *Adv. Genet.* **9**, 217–281.

Whitehouse, H.L.K. 1950 Multiple allelomorph incompatibility of pollen and style in the evolution of angiosperms. *Ann. Bot. N.S.* **14**, 198–216.

Zaki, M.A.M. & Dickinson, H.G. 1990 Structural changes during the first divisions of embryos resulting from anther and free microspore culture in *Brassica napus*. *Protoplasma* **156**, 149–162.

Zaki, M.A.M. & Dickinson, H.G. 1991 Microspore derived embryos in *Brassica*; the significance of division symmetry in pollen mitosis I to embryogenic development. *Sex. Pl. Reprod.* **4**, 48–55.

3

Sex determining gene expression during embryogenesis

R. Lovell-Badge

SUMMARY

The Y-linked gene *Sry* acts during a critical period of gonadal differentiation to divert the normal or default pathway of gene activity that would otherwise lead to the development of ovaries into one that leads to the development of testes. It acts cell autonomously, probably within the cell lineage that gives rise to Sertoli cells in the testis or follicle cells in the ovary. The remaining cell types within the gonad, each of which has a developmental choice, then become fated to follow the testicular pathway. This process must depend on cell–cell interactions as *Sry* is not required within these other cell types for their differentiation. Subsequent male development of the animal as a whole is dependent on the production of testosterone and other factors by the testis.

Sry encodes a DNA binding protein of the HMG box class, and presumably acts to regulate the expression of other genes which then confer cellular phenotype. However, rather than operating like other classes of transcription factor, it has been shown to induce a dramatic bend in its DNA binding sites, and may not directly affect transcription of target genes. Instead, it may permit other factors to interact, which in turn either activate or repress transcription. Sequence comparisons between *Sry* genes from various species suggest that the HMG box is the only functional part of the protein. This part is responsible for DNA binding, and both mouse and human SRY bind the same consensus sequence at high affinity *in vitro*. However, the human gene fails to cause female to male sex reversal in transgenic mice. Possible reasons for this are discussed. There is also much evidence, including transgenic data, to suggest that the level of expression of *Sry* is critical to its function. On top of this, the gene can only successfully induce testis differentiation if the correct cell–cell interactions occur within the developing gonad. Despite knowing the identity of the testis determining gene, we are therefore still a long way from understanding how it achieves its function.

1. INTRODUCTION

For those interested in the mechanisms underlying decisions of cell fate, sex determination provides an obvious subject for study. Most species exhibit an extensive range of dimorphic characteristics that distinguish the two sexes, and the genetic pathways underlying the differences can often be studied with relatively few problems of lethality as mutations tend to lead to transformation of sex. However, it is clear that widely differing mechanisms operate to control sex determination and differentiation among different animal species. This is evident in the type of switch used to initiate either the program of male or of female development. Thus, the switch can be environmental as in many reptiles and fishes, or it may correspond to the number of X chromosomes, as in *Drosophila* and *C. elegans*, or it may be a dominant genetic switch as in mammals where the inheritance of a Y chromosome at fertilization determines maleness (Ford *et al.* 1959; Jacobs & Strong 1959; Welshons & Russell 1959).

There are many other fundamental differences between mammalian sex determination and that of *Drosophila* and *C. elegans*, the two species for which we know the most (Hodgkin 1992). In the latter two species, because sex determination is linked to the X chromosome dosage compensation mechanism, it has to occur early. In *Drosophila*, the pathway appears to act entirely cell autonomously, with many of the steps involving post-transcriptional control of gene activity. In *C. elegans*, it is also initiated in all cells in a cell autonomous manner, although the pathway includes a short range extracellular step, which may serve to ensure that no mistakes are made (Hunter & Wood 1992). In mammals it is not linked to dosage compensation, as the number of X chromosomes is irrelevant to the resulting sex of an individual. Also, even though genetic sex is established at fertilization, the decision as to which pathway to follow operates much later on in development, during the period of organogenesis. This involves the differentiation of the gonadal primordia, the genital ridges, into testes rather than ovaries if the Y-linked testis determining gene is present (see Capel & Lovell-Badge 1993). The male determining signal is then effectively exported, potentially to all cells of the body, through the production of two factors by the testis, anti-Müllerian hormone (AMH, otherwise known as Müllerian inhibiting substance or MIS) and testosterone (Münsterberg & Lovell-Badge 1991; Capel & Lovell-Badge 1993). It is the action of these factors that leads to the dramatic differences between males and females in anatomy, physiology and behaviour. The female pathway can be considered the default pathway, as it will occur in the absence of gonads, whether they are removed from XX or XY embryos (Jost 1947; Jost *et al.* 1973).

The critical event in sex determination in mammals, therefore, is the action of the Y-linked testis determining gene. This switches the pathway of gene activity which leads to the differentiation of ovaries into one which leads to the differentiation of testes. The central problem is to understand how this switch is brought about.

There is now substantial evidence, both circumstantial and direct, to prove that *SRY* (Sex determining region, Y gene) is the testis determining gene. *SRY* was found through a positional cloning strategy which involved looking for conserved genic sequences within the smallest region of the human Y chromosome found in XX males (Sinclair *et al.* 1990). The mouse homologue, *Sry*, was also found to map to the minimum portion of the Y chromosome known to be sex determining in this species (Gubbay *et al.* 1990). In addition there appears to be an *SRY* homologue on the Y chromosome of all mammals with a Y chromosomal sex determining mechanism. This includes metatheria (marsupials) as well as eutheria (Sinclair *et al.* 1990; Foster *et al.* 1992). The gene was found to be expressed at the right time in the right tissue for it to be sex determining (see below) (Koopman *et al.* 1990). It was also found to encode a DNA binding protein consistent both with the presumed regulatory role of the testis determining gene and with the prediction that it acts cell autonomously (Gubbay *et al.* 1990; Sinclair *et al.* 1990). Direct evidence that *SRY/Sry* is the Y encoded testis determining gene comes from two sources. First, the association of mutations within the DNA binding domain of SRY with male to female sex reversal in humans (Harley *et al.* 1992), and secondly, female to male sex reversal found when a small genomic fragment carrying the mouse *Sry* gene was introduced as a transgene into XX embryos (Koopman *et al.* 1991). This latter experiment showed that *Sry* was not only involved in testis determination, but that it is the only gene from the Y chromosome that is required for this process.

The purpose of this report is to discuss data relevant to how *SRY/Sry* might function. It is first necessary to consider how the gonads develop and how *Sry* is expressed.

2. GONADAL DEVELOPMENT AND *SRY* EXPRESSION

The development of the gonads is intimately linked with that of the mesonephros, the primitive kidney, which itself develops from unsegmented, intermediate lateral plate mesoderm (see Capel & Lovell-Badge 1993). The genital ridges arise as a thickening on the ventral–medial surface of each mesonephros, from about 10 dpc in the mouse. This appears to involve the ingression of mesenchymal cells from the coelomic epithelium, or the epithelium of mesonephric or Wolffian ducts. The genital ridges increase in volume but remain indifferent, i.e. morphologically identical between the two sexes, until 11.5 dpc. Between 11.5 and 12.5 dpc the developing testis takes on a characteristic stripey appearance, due to the differentiation and alignment of Sertoli cells into epithelial testis cords. No overt sign of differentiation occurs in the developing ovary until several days later when follicle cells organize around oocytes. However, there may be earlier changes in the state of the cells as there is some evidence to suggest that competence to form a testis exists for only a limited period (Palmer & Burgoyne 1991*c*).

Experiments to detect *Sry* transcripts, mainly using the reverse transcriptase–polymerase chain reaction (RT–PCR) method, but backed up by *in situ* hybridization data, show that the gene is expressed for a limited period between 10.5 and 12.0 dpc (Koopman *et al.* 1990). This expression is in cells of the genital ridge as it develops, with no sign of expression beforehand at 9.5 dpc and only traces sometimes present at 12.5 dpc when testis cord formation is apparent. This expression for a brief period only, suggests that *Sry* initiates testis development and that it is not required for long term maintenance of testis specific gene expression.

Our current working hypothesis assumes that there are four cell lineages that make up the developing mammalian gonad, each of which is effectively bipotential (see, for example, Lovell-Badge 1992). These comprise the germ cells, which migrate into the genital ridge between the tenth and eleventh day of development, and three somatic cell types: (i) the supporting cell precursors, which give rise to Sertoli cells in the male and follicle cells in the female; (ii) steroid-producing cell precursors, which give Leydig cells in the male, theca or interstitial cells in the female; and (iii) connective tissue cells, which include those that give vascular tissue (particularly prominent in testis), tunica and peritubular myoid cells in the testis, and tunica and stromal cells in the ovary.

Testes develop normally in embryos homozygous for mutations at the *W* (or *white spotting*) locus, which compromise the ability of germ cells to proliferate and colonize the genital ridges (Mintz & Russell 1957). The germ cells are therefore irrelevant to testis determination. Expression of *Sry* is unaffected in *W/W* mutant embryos, confirming that *Sry* is expressed by one or more of the somatic cell types (Koopman *et al.* 1990). On the other hand, germ cells are influenced by their environment (McLaren 1985). In the fetal testis they enter mitotic arrest, resuming mitosis and entering meiosis after birth. In an ovary, or in ectopic sites, they enter meiosis and arrest in the first meiotic prophase. In both cases, their differentiation is independent of their sex chromosome constitution.

Of the three somatic cell lineages the supporting cells would appear to be the most critical for testis determination. Burgoyne and colleagues (Burgoyne *et al.* 1988; Palmer & Burgoyne 1991*a*) looked at the sex chromosome constitution of cells within the testes of XX↔XY chimaeras, and found that Sertoli cells were almost exclusively XY, whereas the proportion of XX and XY cells in the other gonadal lineages was the same as that found elsewhere in the embryo. This suggested that the testis determining gene acts cell autonomously within the supporting cell precursors. We assume these are the cells that express *Sry*, and that this expression is responsible for their differentiation along the Sertoli cell pathway. However, recent experiments by Buehr *et al.* (1992) show that 11.5 dpc male genital ridges, isolated from their mesonephros, fail to organize to give testis cords. This suggests additional factors are required, in addition to *Sry* expression, for Sertoli cells to complete their differentiation. When the genital ridges were explanted together with mesonephros, testis cords developed normally. By using mesonephros isolated from embryos carrying a transgene marker, Buehr and colleagues were then able to demonstrate that a population of cells migrate into the developing testis from the mesonephros after 11.5 dpc (i.e. when *Sry* transcription is declining). These give rise to the peritubular myoid cells. These would appear to be essential for the organization of pre-Sertoli cells into testis cords, and therefore for their proper differentiation. This must be due to either a short range paracrine effect or direct cell–cell contact.

The migration of these mesonephros-derived cells, and their differentiation into peritubular myoid cells, must in turn be due to an interaction with the pre-Sertoli cells. Other signals from the differentiating Sertoli cells must also be responsible for the entry of germ cells into mitotic arrest and the differentiation of Leydig cells, etc.

From the above it would appear that the role of *Sry* is not to irreversibly commit supporting cell precursors to the Sertoli cell pathway. Other data from XX↔XY chimaeras or mosaics would agree with this. When the proportion of XY cells is less than about 25% (depending on strain combination) these embryos develop as females. However, perhaps contrary to expectation, there is no bias in the formation of XY follicle cells. These are found in the same proportions as XY cells in other tissues (Palmer & Burgoyne 1991*b*). It would appear that *Sry*-expressing pre-Sertoli cells are required to induce their own differentiation, perhaps by an autocrine effect, and that they fail to do so when present in too low a number. But, from the experiments of Buehr *et al.* (1992), it is more likely that there is a failure to induce myoid cells and in the absence of the return signal from these cells, the Sertoli cell pathway is blocked and the supporting cells revert to the follicle cell pathway.

Sry is also not essential for Sertoli cell differentiation. There are a number of natural or experimental conditions where XX follicle cells take on the appearance and many of the biochemical properties of Sertoli cells. These cases include Freemartin cattle (Jost *et al.* 1975), grafts of foetal mouse ovaries under the kidney capsule of newborns (Taketo-Hosotani 1987), and female mice expressing high ectopic levels of an AMH transgene (Behringer *et al.* 1990). In all cases, the transition to Sertoli cell phenotype is associated with loss of oocytes from the ovaries. This implies a tendency for supporting cells to make Sertoli cells unless prevented from doing so by an oocyte-derived factor. Any such factor could not play a role in the normal sex determination process, as it would be absent from XX embryos lacking germ cells due to mutations at *W*, and these have only a streak gonad without any sign of Sertoli cell differentiation.

3. THE INTERACTION OF SRY PROTEIN WITH DNA

Our thinking has perhaps been constrained by the notion of master regulatory genes acting to direct cellular differentiation down specific pathways. For example, *MyoD* was thought to operate in this way, giving muscle cell differentiation when introduced by itself into a variety of different cell types (Weintraub *et al.* 1991; Rudnicki *et al.* 1992). However, *Sry* can not possibly act in this way. Although it functions as the normal switch to induce Sertoli cell differentiation, in some circumstances it is clearly not sufficient or even necessary. Instead, the gene appears to influence cell fate rather than direct it. It is worth considering the molecular properties of SRY as this may provide clues to its mode of action.

SRY was one of the first members of what is now a very large and still expanding group of DNA binding proteins united by possession of a domain referred to as an HMG box. The latter was first described as a region of homology between the high mobility group non-histone protein HMG1, and hUBF, a cofactor of RNA polymerase 1 required for ribosomal RNA transcription (Jantzen *et al.* 1990). HMG1 is a highly abundant and ubiquitous chromatin associated protein which binds linear DNA with low affinity and without apparent sequence specificity (Bianchi *et al.* 1992). Despite being associated with its target genes, hUBF has no clear consensus DNA sequence motif to which it binds, and its interaction with DNA is fairly weak (Jantzen *et al.* 1990). There are a number of other DNA binding proteins of this sort which can be loosely grouped together by their property of having little or no sequence specificity, being expressed ubiquitously and tending to have more than one HMG box. SRY falls into a second class, however, the members of which often show highly restricted tissue distribution, bind to specific sequences at high affinities and only have a single HMG box (Ner 1992). The two best characterized members of this class, apart from SRY itself, are TCF1 and LEF1 (or TCF1α) which are expressed in lymphoid cells and are involved in transcriptional regulation of, for example, T-cell specific genes such as CD3ε and TCRα (Travis *et al.* 1991; van de Wetering *et al.* 1991).

All these factors interact with DNA through the HMG box which is approximately 79 amino acids in length. Mutations in the HMG box of SRY have been associated with cases of sex reversal in humans leading to XY females (Harley *et al.* 1992). *In vitro* DNA binding studies have also shown that the boxes alone, from hUBF, LEF-1 and SRY, are sufficient for DNA binding to target sequences (Jantzen *et al.* 1990; Giese *et al.* 1991; Nasrin *et al.* 1991). SRY has been shown to bind the same target sequence as TCF1, AACAAAG, but the sequence AACAAT is more likely to be the highest affinity site for both human and mouse SRY (Denny *et al.* 1992; Giese *et al.* 1992; Harley *et al.* 1992, unpublished results).

Grosschedl and colleagues (Giese *et al.* 1992), and more recently Bianchi and coworkers (Ferrari *et al.* 1992), have looked at the interaction of HMG box containing proteins with DNA in more detail. LEF-1 and SRY, and probably other members of this class, induce a dramatic bend in target DNA on binding. This bend is in the order of 130°, which is more than any other known transcription factor, although IHF, the *E. coli* Integration Host Factor, bends DNA by 140°. Also, like IHF, SRY and LEF1 contact DNA in the minor groove, which is unusual compared to known transcription factors (Giese *et al.* 1991; van de Wetering *et al.* 1991).

HMG1 binds to four-way DNA junctions and cruciform DNA structures at much higher affinities than it does to linear DNA. These cruciform structures, which may be made artificially by using four synthetic oligomers, also form at inverted repeats and at Holliday junctions during recombination. They have two angles of 60° and two of 120° and HMG1 is thought to bind to the former, which would be equivalent to a bend of 120° in linear DNA (Bianchi *et al.* 1992). Bianchi's recent results also show that SRY will bind the same cruciform structures as HMG1, without sequence specificity and with affinities as high as its binding with AACAAT motifs in linear DNA (Ferrari *et al.* 1992). Probably all HMG boxes have this ability to bind structured DNA, but the binding of linear DNA, and its subsequent bending, is an extra property of the SRY/LEF1 class.

While SRY appears to bind the same structures as HMG1, it seems unlikely that it would do so within the context of the nucleus. HMG1 is such an abundant protein, and can be assumed to occupy all such sites. SRY is therefore likely to bind only its free specific target sequences in linear DNA. This binding will then induce the dramatic bend. Until target genes are identified it is only possible to speculate how this may affect transcription, but a comparison with LEF1 may be pertinent. Co-transformation studies with LEF-1 and a target gene construct shows that it only has weak activation properties by itself (Giese *et al.* 1992). However, the LEF1 binding site is flanked by sites for two other transcription factors. It appears that the bend in DNA induced by LEF1 binding, brings these two sites together, allowing the other factors to interact. This interaction then activates transcription. The distance between the sites for all three factors is critical, as is the degree of bending induced by LEF1. One could equally imagine schemes by which the bending prevented the binding or interaction of other factors, leading to target gene repression.

HMG box containing factors, such as SRY, could therefore facilitate or prevent the action of other specific transcription factors on target gene expression, while having no direct influence by themselves on RNA polymerase activity. This could provide a molecular explanation for why *Sry* is not sufficient to induce Sertoli cell differentiation during normal testis determination, and why it may not be necessary in other circumstances.

4. TRANSGENIC STUDIES RELEVANT TO SRY ACTION

A 14 kb mouse genomic DNA fragment carrying *Sry* was shown to be sufficient to induce testis formation when introduced as a transgene into chromosomally female embryos (Koopman *et al.* 1991). This provided the formal proof that *Sry* is the Y encoded testis determining gene. Moreover, it showed that it was the only gene from the Y chromosome needed for normal male development. The sex reversed mice were sterile, but this is a separate problem, due to the presence of two X chromosomes which are incompatible with spermatogenesis and to the lack of genes involved in this process which are known to map elsewhere on the Y chromosome.

This 14 kb genomic fragment failed to cause sex reversal in all cases. In fact only about 25% of XX transgenic mice develop as males. In some cases this is likely to be due to mosaicism for the presence of the transgene, due to integration after the first cleavage division, such that it is present in less than 25% of supporting cell precursors. Alternatively, it is known that transgenes often show very different levels of expression from one line to another, probably due to the effect of genomic sequences at the site of integration. Since the purpose of our experiments was to assay function and not simply tissue specific expression, the timing or level of expression of the *Sry* transgene may be ineffective in some cases. Evidence for this comes from the study of one particular transgenic line, termed 32.10 (N. Vivian, P. Koopman & R. Lovell-Badge, unpublished data). The founder animal was a fertile female who transmitted the transgene to both normal daughters and sons. This ruled out mosaicism as a possible reason for the failure to sex reverse. However, subsequent breeding of these animals has led to instances of sex reversal, either when they were bred with wild-type (one case) or with other transgenic siblings (five cases). Two intersex offspring have also been obtained from these matings. Preliminary results showed that *Sry* is being expressed in the genital ridge of transgenic XX embryos at 11.5 dpc. Careful measurement of the number of transcripts over a time course of development is required, but the expression of the transgene appeared to be at a lower level than normal. It is likely that *Sry* expression is at a threshold level, and homozygosity for the transgene increases the incidence of sex reversal. However, not all homozygotes are male and some hemizygotes are, suggesting that genetic background effects segregating in these F2(C57BL/10 × CBA) and later generation crosses are affecting the ability of the transgene to induce testis differentiation.

Transgenic mice have also been generated using genomic fragments carrying the human *SRY* gene (Koopman *et al.* 1991; Palmiter, personal communication). However, no evidence of sex reversal has been seen despite apparently normal expression of the transgene. No difference has been seen so far in the interaction of mouse and human SRY protein with DNA *in vitro*, both bind the same target DNA sequences, and appear to induce similar degrees of bending. Why then does the human gene not work in mice?

If the simple act of bending the DNA is sufficient to cause an effect on transcription then there is no need to have other domains associated with the HMG box for activation or interaction with other proteins. This is clearly not the case for many of the HMG box containing genes, as they have such putative sequences which are well conserved amongst different species. However, when mouse and human *Sry/SRY* sequences are compared there is no homology outside the HMG box. This lack of homology is, in fact, evident when comparing SRY from any two species. On the other hand, although the HMG box clearly represents a conserved domain between SRY proteins from different species, the degree of homology within the box is much less than that found when comparing other HMG box containing proteins. Indeed, if all the known SRY sequences are compared with those of *Sry*-related HMG box containing genes (these so-called *Sox* genes are located on chromosomes other than the Y), no particular amino acid residues are found that are characteristic of the SRY proteins (Foster *et al.* 1992). There are several possibilities to explain the transgenic results: (i) the variation within the HMG box is significant and the HMG boxes of human and mouse SRY interact with target sequences in subtly different ways not detectable *in vitro*; (ii) it is significant, but affects interaction with other proteins and not DNA; or (iii) this variation is not significant, and the failure of the human gene to work in mice is due to the variation in the non-HMG box parts of the protein. Whichever possibility applies it suggests that there is very rapid evolution of protein–DNA or protein–protein interactions that can affect SRY action. Carrying out domain swap experiments between the human and mouse genes in transgenic mice should help to resolve this problem.

5. CONCLUSIONS

Sry clearly acts as the normal switch in mammalian sex determination. However, we do not know how it affects the expression of other genes required for Sertoli cell differentiation. Parallels drawn between SRY and other HMG box containing proteins suggest that it may be a cofactor of transcription that mediates the interaction of other transcription factors. Alternatively, it may alter the accessibility of these other factors to target sites within chromatin. However, we need to identify the target genes. This is difficult, and at present we have no direct information to indicate whether SRY is an activator of genes involved in testis determination, a repressor of genes involved in the ovarian determination pathway, or both.

It is perhaps easier to consider SRY acting as a repressor. The lack of conservation between species suggests that there are few interactions with other proteins, and, simply by bending DNA, SRY could prevent the binding of other transcription factors (but see the discussion above). Some genetic data are also easier to reconcile with this possibility. A significant proportion of human XX males do not have any Y chromosome sequences, let alone *SRY* (Palmer *et al.* 1989). The frequency of these is greater than expected for gain of function mutations, but would be appropriate for loss of function mutations in a downstream gene if SRY is a repressor of this gene. If *Sry* is a repressor, it must act on the female pathway just long enough to allow the male pathway of gene activity to operate. If so, this subtly alters our notion of the former being the default pathway.

REFERENCES

Behringer, R.R., Cate, R.L., Froelick, G.J., Palmiter, R.D. & Brinster, R.L. 1990 Abnormal sexual development in transgenic mice chronically expressing Müllerian inhibiting substance. *Nature, Lond.* **345**, 167–170.

Bianchi, M.E., Falciola, L., Ferrari, S. & Lilley, D.M.J. 1992 The DNA binding site of HMG1 protein is composed of two similar segments (HMG boxes), both of which have counterparts in other eukaryotic regulatory proteins. *EMBO J.* **11**, 1055–1063.

Buehr, M., Gu, S. & McLaren, A. 1992 Mesonephric contribution to testis differentiation in the fetal mouse. *Development* **117**. (In the press.)

Burgoyne, P.S., Buehr, M., Koopman, P., Rossant, J. & McLaren, A. 1988 Cell autonomous action of the testis-determining gene: Sertoli cells are exclusively XY in XX XY chimaeric mouse testes. *Development* **102**, 443–450.

Capel, B. & Lovell-Badge, R. 1993 The Sry gene and sex determination in mammals. *Adv. Devl Biol.* **2**. (In the press.)

Denny, P., Swift, S., Connor, F. & Ashworth, A. 1992 A testis specific gene related to SRY encodes a sequence-specific DNA binding protein. *EMBO J.* **10**, 3705–3712.

Ferrari, S., Harley, V.R., Pontiggia, A., Goodfellow, P.N., Lovell-Badge, R. & Bianchi, M.E. 1992 SRY, like HMG1, recognizes sharp angles in DNA. *EMBO J.* **11**, 4497–4506.

Ford, C.E., Jones, K.W., Polani, P.E., de Almeida, J.C. & Briggs, J.H. 1959 A sex-chromosome anomaly in a case of gonadal dysgenesis (Turner's syndrome). *Lancet* **i**, 711–713.

Foster, J.W., Brennan, F.E., Hampikian, G.K., Goodfellow, P.N., Sinclair, A.H., Lovell-Badge, R., Selwood, L., Renfree, M.B., Cooper, D.W. & Marshall Graves, J.A. 1992 Evolution of sex determination and the Y chromosome: SRY-related sequences in marsupials. *Nature, Lond.* **359**, 531–533.

Giese, K., Amsterdam, A. & Grosschedl, R. 1991 DNA-binding properties of the HMG domain of the lymphoid-specific transcriptional regulator LEF-1. *Genes Dev.* **5**, 2567–2578.

Giese, K., Cox, J. & Grosschedl, R. 1992 The HMG domain of lymphoid enhancer factor 1 bends DNA and facilitates assembly of functional nucleoprotein structures. *Cell* **69**, 1–20.

Gubbay, J., Collignon, J., Koopman, P., Capel, B., Economou, A., Münsterberg, A., Vivian, N., Goodfellow, P. & Lovell-Badge, R. 1990 A gene mapping to the sex-determining region of the mouse Y chromosome is a member of a novel family of embryonically expressed genes. *Nature, Lond.* **346**, 245–250.

Harley, V.R., Jackson, D.I., Hextal, P.J., Hawkins, J.R., Berkovitz, G.D., Sockanathan, S., Lovell-Badge, R. & Goodfellow, P.N. 1992 DNA binding activity of recombinant SRY from normal males and XY females. *Science, Wash.* **225**, 453–456.

Hodgkin, J. 1992 Genetic sex determination, mechanisms and evolution. *Bioessays* **14**, 253–261.

Hunter, C.P. & Wood, W.B. 1992 Cell interactions in C. elegans sex determination: genetic evidence from mosaic analysis of the masculinizing gene her-1. *Nature, Lond.* **355**, 551–555.

Jacobs, P.A. & Strong, J.A. 1959 A case of human intersexuality having a possible XXY sex-determining mechanism. *Nature, Lond.* **183**, 302–303.

Jantzen, H.-M., Admon, A., Bell, S.P. & Tjian, R. 1990 Nucleolar transcription factor hUBF contains a DNA-binding motif with homology to HMG proteins. *Nature, Lond.* **44**, 830–836.

Jost, A. 1947 Recherches sur la différenciation sexuelle de l'embryon de lapin. Archs *Anat. microsc. Morph. Exp.* **36**, 271–315.

Jost, A., Vigier, B., Prépin, J. & Perchellet, J. P. 1973 Studies on sex differentiation in mammals. *Recent Prog. Horm. Res.* **29**, 1–41.

Jost, A., Perchellet, J.P., Prepin, J. & Vigier, B. 1975 The prenatal development of bovine freemartins. In *Symposium on intersexuality* (ed. R. Reinborn), pp. 392–406. Berlin: Springer-Verlag.

Koopman, P., Münsterberg, A., Capel, B., Vivian, N. & Lovell-Badge, R. 1990 Expression of a candidate sex-determining gene during mouse testis differentiation. *Nature, Lond.* **348**, 450–452.

Koopman, P., Gubbay, J., Vivian, N., Goodfellow, P. & Lovell-Badge, R. 1991 Male development of chromosomally female mice transgenic for *Sry*. *Nature, Lond.* **351**, 117–121.

Lovell-Badge, R. 1992 The role of Sry in mammalian sex determination. In *Postimplantation development in the mouse* (Ciba Foundation Symposium **165**), pp. 162–182. Chichester: Wiley.

McLaren, A. 1985 Relating of germ cell sex to gonadal differentiation. In *The origin and evolution of sex* (ed. H. O. Halvorsen & A. Monroy), pp. 289–300. New York: Liss.

Mintz, B. & Russell, E.S. 1957 Gene-Induced embryological modifications of primordial germ cells in the mouse. *J. exp. Zool.* **134**, 207–237.

Munsterberg, A. & Lovell-Badge, R. 1991 Expression of the mouse anti-mullerian hormone gene suggests a role in both male amd female sexual differentiation. *Development* **113**, 613–624.

Nasrin, N., Buggs, C., Kong, X.F., Carnazza, J., Goebl, M. & Alexander-Bridges, M. 1991 DNA-binding properties of the product of the testis-determining gene and a related protein. *Nature, Lond.* **354**, 317–320.

Ner, S.S. 1992 HMGs everywhere. *Curr. Biol.* **2**, 208–210.

Palmer, M.S., Sinclair, A.H., Berta, P., Ellis, N.A., Goodfellow, P.N., Abbas, N.E. & Fellous, M. 1989 Genetic evidence that ZFY is not the testis-determining factor. *Nature, Lond.* **342**, 937–939.

Palmer, S.J. & Burgoyne, P.S. 1991*a* In situ analysis of fetal, prepuberal and adult XX-XY chimaeric mouse testes: Sertoli cells are predominantly, but not exclusively, XY. *Development* **112**, 265–268.

Palmer, S. & Burgoyne, P.S. 1991*b* XY follicle cells in the ovaries of XO/XY and XO/XY/XYY mosaic mice. *Development* **111**, 1017–1020.

Palmer, S.J. & Burgoyne, P.S. 1991*c* The *Mus musculus domesticus Tdy* allele acts later than the *Mus musculus musculus Tdy* allele: a basis for XY sex reversal in C57BL/6-Ypos mice. *Development* **113**, 709–714.

Rudnicki, M.A., Braun, T., Hinuma, S. & Jaenisch, R. 1992 Inactivation of MyoD in mice leads to up-regulation of the myogenic HLH gene Myf-5 and results in apparently normal muscle development. *Cell* **71**, 383–390.

Sinclair, A.H., Berta, P., Palmer, M.S., Hawkins, J.R., Griffiths, B.L., Smith, M.J., Foster, J.W., Frischauf, A.-M., Lovell-Badge, R. & Goodfellow, P.N. 1990 A gene from the human sex-determining region encodes a protein with homology to a conserved DNA-binding motif. *Nature, Lond.* **346**, 240–244.

Taketo-Hosotani, T. 1987 Factors involved in the testicular development from fetal mouse ovaries following transplantation. *J. exp. Zool.* **241**, 95–100.

Travis, A., Amsterdam, A., Belanger, C. & Grosschedl, R. 1991 LEF-1, a gene encoding a lymphoid-specific protein with an HMG domain, regulates T-cell receptor alpha enhancer function. *Genes Dev.* **5**, 880–894.

Weintraub, H., Davis, R., Tapscott, S., Thayer, M., Krause, M., Benezra, B., Blackwell, T.K., Turner, D., Rupp, R., Hollenberg, S., Zhuang, Y. & Lassar, A. 1991 The myoD gene family: nodal point during specification of the muscle cell lineage. *Science, Wash.* **251**, 761–766.

Welshons, W.J. & Russell, L.B. 1959 The Y chromosome as the bearer of male determining factors in the mouse. *Proc. natn. Acad. Sci. U.S.A.* **45**, 560–566.

van de Wetering, M., Oosterwegel, M., Dooijes, D. & Clevers, H. 1991 Identification and cloning of TCF-1, a T lymphocyte-specific transcription factor containing a sequence-specific HMG box. *EMBO J.* **10**, 123–132.

4

The inheritance of germline-specific epigenetic modifications during development

M. A. Surani, H. Sasaki, A. C. Ferguson-Smith, N. D. Allen, S. C. Barton, P. A. Jones and W. Reik

SUMMARY

Parental genomes in mammals are programmed in the germline with heritable epigenetic modifications that exert control on the expression of specific (imprinted) genes. DNA methylation is one form of epigenetic modification which shows marked genome-wide variations in the germline and during early development. Certain transgene loci also demonstrate (reversible) germline-specific methylation imprints that are heritable in somatic tissues during development. Recently, four endogenous genes have been identified whose expression is dependent on their parental origin. The mechanism of genomic imprinting and the role of imprinted genes during development is beginning to be analysed. Three of these genes map to the mouse chromosome 7. Human chromosomes 11p13, 11p15, and 15q11–13 are associated with disorders exhibiting parental origin effects in their patterns of inheritance. These regions share syntenic homology with mouse chromosome 7. The relationship between parental imprints, germline-dependent epigenetic inheritance and totipotency is also under investigation using embryonic stem cells derived from the epiblast. These cells are pluripotent or totipotent and evidence indicates the presence of at least the primary parental imprints. However, imprints inherited from the paternal germline in androgenetic cells are apparently more stable than those from the female germline in parthenogenetic cells.

1. INTRODUCTION

At fertilization, each parent contributes a genetically equivalent haploid set of autosomes to the zygote. Although most parental loci function identically during development, there are exceptions to this rule. Expression of some autosomal loci is in fact determined by whether they are of maternal or paternal origin (Cattanach & Beechey 1990). This 'memory' of parental origin of chromosomes controlling gene expression must require heritable epigenetic modifications. It is crucial that the germline must also be able to reverse any previously acquired epigenetic modifications. This whole phenomenon is called genomic imprinting. The non-reciprocal expression of some parental alleles plays a role in development and aberrant expression results in abnormal phenotypes and certain genetic disorders in the human (Solter 1988; Surani & Reik 1922). The molecular mechanism of genomic imprinting is so far scarcely understood. The initial studies were prompted by the identification of some transgene inserts which showed parental origin effects in their epigenetic inheritance and expression (Reik *et al.* 1990; Surani *et al.* 1990). These studies can now be extended to the recently identified endogenous imprinted genes.

Parental origin-specific imprinting probably occurs progressively in which both the germline-specific and post-zygotic events play a role in the process (Surani 1991). The germline-dependent modifications are simplest to envisage as the homologous chromosomes are segregated and subject to distinct influences during spermatogenesis and oogenesis. However, the parental alleles could be subjected to further differential modifications after fertilization both before and after pronuclear fusion. At this time nuclear-cytoplasmic interaction and oocyte cytoplasmic factors may have a significant part to play in epigenetic modifications. (Klose & Reik 1992). Some of the major epigenetic changes could therefore occur during pre-implantation development at a time when the pluripotent–totipotent epiblast cells are established. Thus, the epiblast cells most probably contain the primary parental imprints inherited from sperm and oocytes at fertilisation; after implantation and as gastrulation commences, further epigenetic modifications could occur in response to these signals coupled with conversion to more stable parental imprints.

There are two observations that implicate a role for DNA methylation in the events involved in imprinting. The first of these is the genome-wide changes in DNA methylation in the germline and early embryos (Bird 1986; Sanford *et al.* 1987; Monk 1988; Reik *et al.* 1990; Kafri 1992). The second concerns the methylation imprinting of certain transgenic loci (Reik *et al.* 1990; Surani *et al.* 1990). CpG dinucleotide methylation is an epigenetic modification known to be associated in some instances, with gene inactivity. For example, X-linked genes containing CpG rich promoter sequences show methylation of these sites on the inactive-X in females (Grant & Chapman 1988). In addition, a crucial role for DNA methylation during embryogenesis has been shown. Mice with a homozygous mutation inactivating the DNA methyltransferase gene are embryonic lethal after implantation (Li *et al.* 1992). Perhaps this is due to an instability of the primary imprints in the absence of methylation. Nevertheless, it is important to note that there are other forms of epigenetic modifications capable of controlling gene expression that are heritable and affect chromatin structure.

Some of the major genome-wide changes in DNA methylation that occur in the germline and early development are worth examining as this provides a background against which we can examine methylation imprinting of transgenes and endogenous genes (Jahner & Jaenisch 1984; Sanford *et al.* 1987; Monk 1988; Kafri 1922). Figure 1 depicts the general genome-wide changes in DNA methylation that have been investigated with respect to both the multi-copy genes as well as single sites within a number of genes. Hence we note dynamic reversible changes in DNA methylation that may have a bearing on parental imprinting of particular loci.

2. PARENTAL IMPRINTING OF TRANSGENES

The initial studies on imprinting were carried out on randomly integrated transgenes which serve as molecular probes for these loci (Reik *et al.* 1990; Surani *et al.* 1990). Two of these transgene inserts are particularly interesting as they show germline-dependent modifications. The MPA434 (Sasaki *et al.* 1992) and RSV-myc (Chaillet *et al.* 1991; Chaillet 1992) both show methylation occurring during oogenesis, i.e. the maternally inherited loci are methylated while the paternal remain hypomethylated (figure 2). There are however some differences between the two transgenes; the paternally inherited RSV-myc transgene undergoes some methylation during early embryogenesis while the *MPA343* transgene remains unmethylated after fertilization. The important point to note is that both transgenes are inherited in their methylated states from the maternal germline and remain so

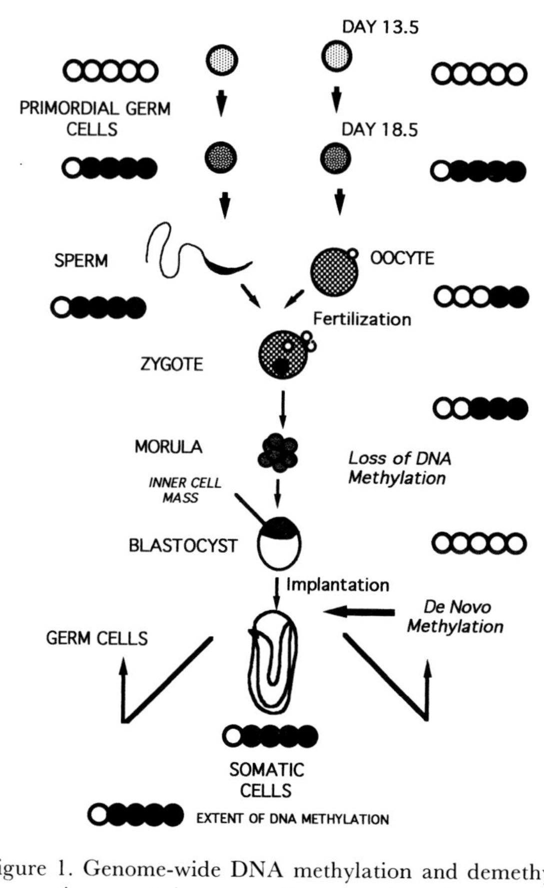

Figure 1. Genome-wide DNA methylation and demethylation events in germ cells and early embryos.

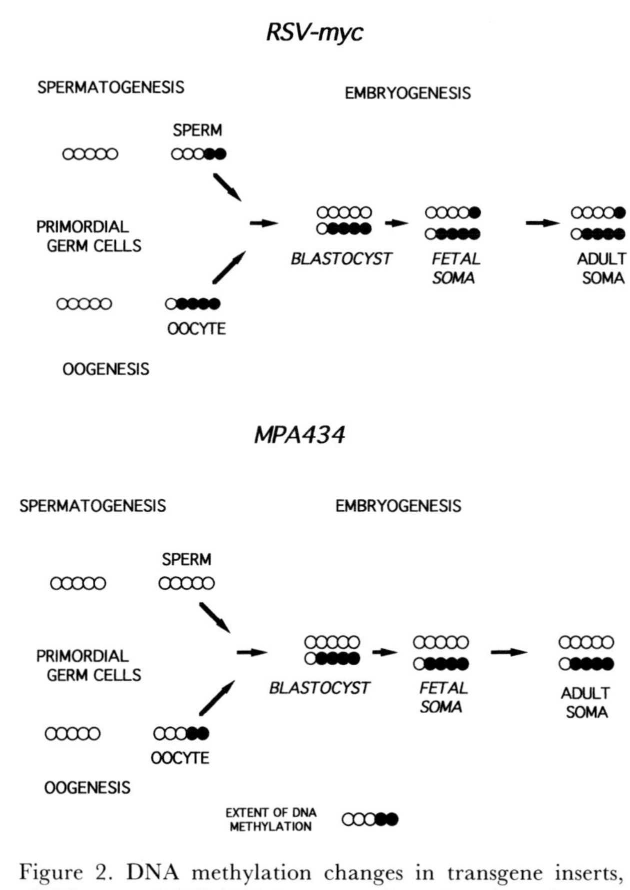

Figure 2. DNA methylation changes in transgene inserts, *RSV-myc* and *MPA434*, in germ cells and early embryos.

thereafter during development. These transgenes probably undergo re-programming in the germline and become demethylated before the initiation of a new cycle of germline-specific parental imprinting. In all the studies reported so far, there are no definitive examples of preferential methylation of transgenes after paternal transmission; it is not clear whether this maternal specific modification to exogenous sequences has any relevance to the imprinting endogenous genes.

Transgene methylation is also evident after fertilization and during early development and this event is often prone to differences in the genetic background of the embryos. One example of this phenomenon was observed with the *TKZ751* locus as a result of position effect (Allen *et al.* 1990). This transgene is methylated on the BALB/c background but remains unmethylated on the DBA background. Strikingly, *TKZ751* transgene methylation requires a maternally inherited BALB/c modifier, i.e. the modifier itself behaves as an imprinted gene. This modifier, which has not yet been identified, is inherited with the nuclear component and is not present in the oocyte cytoplasm. However, the transgene locus *pHRD*, is subject to methylation in a position-independent manner; *pHRD* remains unmethylated and expressed on DBA/2 and SJL genetic background but is methylated with the loss of expression whenever it is crossed in to C57BL/6 mice. (Engler *et al.* 1991). The modification is not affected by the parental origin of the modifier. The modifier locus in this instance has been mapped to the distal region of chromosome 4 and has been called *Ssm-1*. (Engler *et al.* 1992).

It is not clear at present what factors are responsible for the imprinting of these transgene loci. In most instances it is supposed that imprinting occurs as a result of position-effects; i.e. effects would occur as a result of transgene integration into genes or regions susceptible to imprinting modification (Surani *et al.* 1990). One locus that has been examined in detail is the *MPA434* locus. This locus maps to chromosome 11, an imprinted chromosomal region. However, the endogenous sequences at the site of transgene integration do not show parent of origin-dependent DNA methylation. Nevertheless, some sites showed allele specific methylation that were not affected by their parental origin (Sasaki *et al.* 1992). It seems likely therefore that the imprinting of this transgene locus is due to a combination of transgene sequences and host sequences at the site of integration and may not be a good model for the parental imprinting of endogenous sequences.

As discussed above, two transgenes have so far been described that show position-independent patterns of DNA methylation, *pHRD* and *RSV-myc*. Both these transgenes become methylated independent of their site of integration. This implies presence of DNA sequences within those two different transgenes that act as the targets for methylation. So far it is not known what these sequences are or how the imprinting influence is manifested. The nature of these sequences that act as signals for imprinting would be informative, particularly in the context of imprinting of endogenous genes.

3. IMPRINTING OF ENDOGENOUS GENES

Based on non-complementation tests, genetic evidence shows that a number of chromosomal regions must contain imprinted genes. The mouse chromosome 7 is one of the most important in this respect since both a maternal and a paternal copy of this chromosome is essential for normal development (Searle & Beechey 1990). Three imprinted genes have now been identified on this chromosome. Two closely linked, reciprocally imprinted genes, *H19* and *Igf2*, are detected on the distal region of chromosome 7 (De Chiara *et al.* 1992; Bartolomei *et al.* 1991; Ferguson-Smith *et al.* 1991). The third gene, *snrpn*, is more proximal and located near the albino locus (Barr *et al.* 1992).

The maternal duplication or paternal deficiency of distal chromosome 7 resulting from intercrosses involving the T9H translocation is embryonic lethal at around day 16 of gestation (Searle & Beechey 1990; Cattanach & Beechey 1990). These embryos are growth retarded. The paternal duplication or maternal deficiency is also lethal but at an unknown stage that occurs after day 6 although the precise stage and the phenotype are unknown owing to the lack of adequate early markers. However, cells with the paternal duplication with corresponding maternal deficiency can be rescued in chimeras with normal cells (Ferguson-Smith *et al.* 1991). Paternal disomy cells which were apparently uniformly distributed, induced a growth enhancement reciprocal to that seen in non chimaeric maternal disomy embryos. It appears that neither of the parental duplications show any effects on particular cell lineages in contrast to phenotypes associated with parental duplications of the whole genome.

The effect on growth observed in these studies is undoubtedly due to the parental imprinting of the *Igf2* gene; this gene is an embryonal mitogen with an important role during embryogenesis (De Chiara *et al.* 1990). The role of *H19* gene is unknown at present with no evidence to suggest that it encodes a protein in the embryo (Bartolomei & Tilghman 1992). Four mammalian species have been studied regarding the DNA sequences of *H19* but so far no common open reading frame is evident. The *H19* RNA is apparently located in 28*S* particle and not with the translational apparatus (Bartolomei & Tilghman, 1992).

The embryos with maternal duplication and paternal deficiency of chromosome 7 (maternal disomic embryos; MatDi7) provide a powerful experimental system with which to explore the mechanism of imprinting. We have demonstrated clearly that these embryos have twice the levels of H19 compared to the controls since these embryos contain two active maternal alleles (H. Sasaki & M. A. Surani, unpublished results). By contrast the levels of Igf2 transcripts in MatDi7 embryos are negligible (Ferguson-Smith *et al.* 1991). Excess levels of H19 by itself would be detrimental because attempts to produce *H19* gain of function transgenic embryos resulted in embryonic lethality at around day 15 of gestation (Brunkow & Tilghman 1991). The other important points to note regarding the imprinting of *H19* and *Igf2* are that the genes are very closely linked, reciprocally imprinted and display remarkably similar spatial and temporal patterns of expression during development (Bartolomei & Tilghman 1992; Sasaki *et al.* 1992). Taken together, these findings may suggest a functional relationship between the two genes regarding the mechanism resulting in their imprinting and their role in embryogenesis.

4. MECHANISM OF IMPRINTING OF ENDOGENOUS GENES

Little is known about the identity of the heritable epigenetic modifications resulting in the parent of origin specific regulation of endogenous imprinted genes. The temporal sequence of events resulting in a stable and heritable imprint must also be considered. Indeed it is not clear if each endogenous (imprinted) gene is imprinted by the same mechanism. In light of the transgene data, DNA methylation analyses may be informative and can be carried out relatively easily on endogenous imprinted genes by comparing the methylation status of the two parental alleles in normal and MatDi 7 embryos; the latter isolated prior to their death at day 16 of gestation.

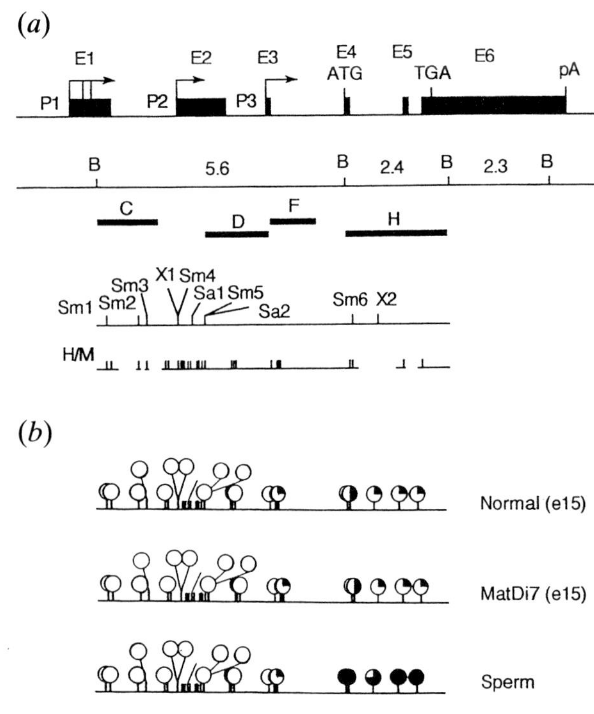

Figure 3. An overall structure and restriction map of the mouse *Igf2* gene (*a*) and summary of methylation analysis (*b*). The positions of six exons (E1–6), three promoters (P1–3) and transcription initiation site (arrows) are indicated. ATG, TGA and pA denote initiation codon, termination codon and major polyadenylation site, respectively. Solid lines indicate the probes which were used in the methylation assays. The methylation-sensitive enzymes are: Sm, *Sma*I; X, *Xho*I; Sa, *Sal*I; H/M, *Hpa*II/*Msp*I. (*b*) The methylation-sensitive enzyme sites are shown by an open circle if unmethylated or a filled one if methylated. Other circles denote intermediate levels of methylation. The methylation status of sites without circles was not determined.

(*a*) *Methylation imprinting of Igf2*

We carried out a detailed analysis of DNA methylation and chromatin structural changes in the *Igf2* gene (Sasaki *et al.* 1992). Two *BamHI* fragments of 5.6 and 2.4 kb were subcloned and used as hybridization probes (figure 3). These fragments contain two strong promoters P2 and P3 and the polypeptide coding exons. P2 contains a CpG island. DNA was isolated from normal and MatDi 7 embryos and was digested with *BamHI* and methylation sensitive enzymes, *HpaII*, *SmaI*, *XhoI* or *SalI*. Any differences in methylation pattern could thus be attributed to the paternal allele absent in MatDi but present in embryos. No detectable differences in methylation were found between control and maternal disomy 7 embryos. The *BamHI* 5.6 kb region was strikingly unmethylated on both alleles suggesting that unlike the imprinting of transgenes, there is no widespread methylation of the maternally repressed *Igf2* gene. However, we cannot exclude the possibility that more subtle differences in DNA methylation exist which are not revealed by analysis using methylation sensitive restriction enzymes alone. The downstream *BamHI* 2.4 kb region however produced a pattern revealing partial methylation in DNA derived from the control and experimental embryos. But again no differences were observed. However, sperm DNA gave a fully methylated pattern in this region.

We have also examined the nuclease sensitivity of the chromatin of the repressed allele in nuclei from the control and MatDi disomic embryos. The 5.6 kb region was found to be highly sensitive to *MspI* in both MatDi7 and control embryos. Similarly DNAaseI hypersensitive sites in MatDi7 and control embryos were examined. Again no differences were found between materials from the two types of embryos. Furthermore prominent DNAase I hypersensitive sites were present in the promoter regions of both alleles. This implies that though one allele is active and the other repressed, both alleles seem equally accessible to transcription factors. Hence, there are no major detectable differences either in DNA methylation or chromatin structure in the active and the repressed loci. This raises the possibility that the chromatin of the repressed maternal allele is potentially active, which was indeed found to be the case when transcription was examined by a sensitive quantitative RT-PCR assay. Low but significant levels of Igf2 transcripts were detected in the MatDi7 embryos. This expression was not attributed to absence of imprinting in a subset of cells as judged by *in situ* hybridization studies for *Igf2* expression in MatDi7 embryos. We conclude that the low level transcription is from most of the cells and not from a subset of cells.

This data raised the possibility that methylation differences may exist elsewhere in the locus. Indeed parent of origin dependent methylation differences were found in a region 3 kb upstream of the first promoter (R. Chaillet, unpublished; Sasaki *et al.* 1992). Multiple *HpaII* sites in this region are slightly more methylated in normal embryos than in MatDi7 embryos showing that the paternal chromosome with the active *Igf2* gene has higher levels of DNA methylation in this region. However, further studies are needed to elucidate the possible significance of this finding for the regulation of *Igf2* by imprinting.

These results for the *Igf2* gene provide no obvious indications for a role of DNA methylation in the imprinting of this locus. Whatever mechanism causes the repression of the *Igf2* gene, it is essential that this should account for the repression from all the three promoters unless evidence is found for promoter specific imprinting of this gene. In addition, the mechanism must not be able to act in the choroid plexus and leptomeninges of the brain where the gene has been shown to escape imprinting (De Chiara *et al.* 1991). We can also rule out the role of major chromatin structural difference such as heterochromatinization in the repression of this gene. Hence, it is necessary to explore more complex possibilities for the imprinting of the *Igf2* gene. Among these possibilities we have to consider if the presence of the closelv linked and reciprocally imprinted *H19* gene is mechanistically relevant for this purpose.

(*b*) *Methylation imprinting of the H19 gene*

H19 gene is approximately 3 kb in size with 5 exons, a defined promoter and two downstream enhancers (Bartolomei & Tilghman 1992). The gene is transcribed by RNA polymerase II spliced and polyadenylated but so far no protein product for this gene has been found in embryos even though it is expressed abundantly. The function of this gene is not known. The promoter lies within a short CpG rich region of approximately 300 b.p. There are other CpG sites which are scattered throughout the locus.

We have commenced DNA methylation analysis of this gene once again by comparing MatDi7 (with the two active *H19* genes) and control embryos (containing the repressed paternal copy of the gene) (A. C. Ferguson-Smith, H. Sasaki & M. A. Surani, unpublished results). The results clearly demonstrate that there are multiple methylatable sites in the promoter region which are methylated on the paternal gene but which are unmethylated on the maternal gene. Examination of the methylation status of these sites in sperm shows that these sites are not methylated. The combined results therefore indicate that methylation occurs after fertilization. There is variable methylation throughout the body of the gene. Further studies have been carried out to examine chromatin structural differences by examining the nuclease accessibility of the promotor and enhancer regions. These studies clearly reveal that there are differences between the active and the repressed loci in the promoter, with the active locus being more accessible (H. Sasaki, A. C. Ferguson-Smith & M. A. Surani, unpublished results).

The methylation of the repressed paternal *H19* gene is therefore more in line with the expectations from imprinting of transgenes and consistent with the inverse correlation between methylation and gene expression (Allen *et al.* 1990; Reik *et al.* 1990; Surani *et al.* 1990; Sasaki *et al.* 1992). However, the results raise some important questions regarding the exact mechanism of imprinting and in particular the nature of the primary imprint. The paternal gene does not appear to inherit its methylation imprint from the sperm but acquires it after fertilization. The precise time when this occurs is unknown at present. However, one possibility is that this methylation occurs after implantation at day 6.5 of gestation when a major *de novo* methylation event is known to occur (Sanford *et al.* 1987; Monk 1990; Li *et al.* 1992). However, the possibility of inheritance of subtle methylation differences by the parental alleles is not excluded nor is it known to what extent the postzygotic methylation differences observed in the case of *H19* will apply in the case of other endogenous imprinted genes. Alternatively, other forms of epigenetic differences between the maternal and paternal copies of the gene may exist from the germline which are recognized subsequently and which result in differential methylation. Further studies are needed to determine the exact nature of this primary imprint.

To address this question though, information on the precise timing of DNA methylation in the promoter region of the paternal allele is required. In particular, it will be highly informative to establish if the *H19* gene is methylated in oocytes. Alternatively, the maternal gene may become methylated during oogenesis but undergoes demethylation during the course of early preimplantation development when genome wide demethylation events are occurring. (see figure 1)

(*c*) *Comparisons between the imprinting of Igf2 and H19*

At the available levels of analysis, there appear to be major differences in the role of DNA methylation in the imprinting of *H19* and *Igf2* (figure 4). One

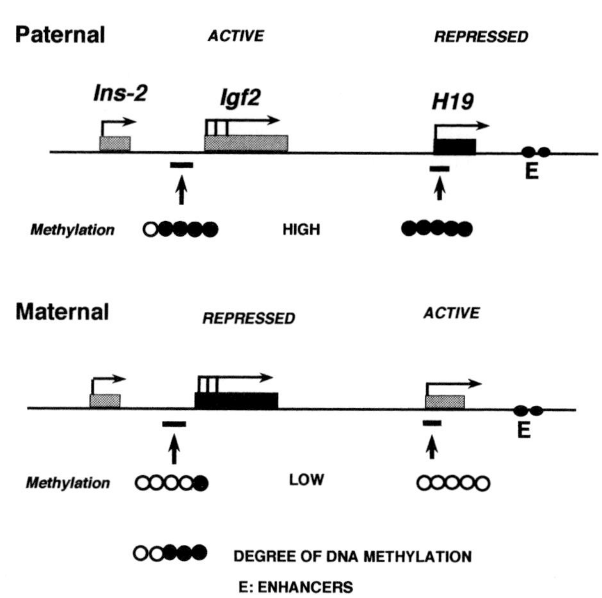

Figure 4. A summary of major parent of origin-dependent methylation difference detected so far at the *Igf2* and *H19* loci.

obvious explanation may be that repression of maternally and paternally inherited genes may require different mechanisms. This question will be resolved when more is known about the imprinting of other endogenous genes such as *Igf2r* on chromosome 17 which is also repressed when paternally inherited (Barlow *et al.* 1991). It will be of interest to determine whether the maternally repressed *Snrpn* gene will share the absence of modification seen with *Igf2*. As far as *Igf2* and *H19* are concerned, one obvious question which arises is the significance, if any, of their close proximity and reciprocal imprinting. The observation of methylation differences in the case of *H19* and a lack of major differences in *Igf2*, prompts speculation on whether *H19* carries the primary imprint while imprinting of *Igf2* occurs secondarily and reciprocally as a consequence of *H19* imprinting. Further studies are required to resolve this and a number of theories addressing this question are discussed elsewhere (Sasaki *et al.* 1992; Bartolomei & Tilghman 1992).

5. IMPRINTING IN ANDROGENETIC AND PARTHENOGENETIC EMBRYOS

The biparental androgenetic and parthenogenetic or gynogenetic embryos produce characteristic phenotypes after implantation; the former with well developed extrembryonic tissues and the latter with relatively well developed embryos (Solter 1988; Surani *et al.* 1990). These reciprocal phenotypes have been suggested to arise as a result of differential expression of imprinted genes. However, no systematic studies have so far been reported to show if the expression of the recently identified imprinted genes occurs appropriately in these embryos. Such studies are important since it would provide proof that the phenotypes are due to the reciprocal expression of imprinted genes. For example, analysis of the 'appropriate' activity of imprinted genes in androgenetic and parthenogenetic conceptuses should also provide information about any interactive roles between the parental genomes in the regulation of imprinting. Studies arising from experiments on the *TKZ751* transgene locus suggest that methylation imprinting segregates with a modifier. Specifically, the modifier from the BALB/c strain of mice induces hypermethylation of the locus and results in the extinction of expression. Evidence suggests that this modifier must be of maternal origin for it to have an effect. In other words the modifier itself reveals a parent of origin effect (Allen & Mooslehner 1992).

It is possible that similar modifiers are important during postzygotic aspects of the imprinting of endogonous genes. For example, it will be interesting to determine methylation status of the *H19* gene in androgenetic embryos in the absence of a maternal genome, since such methylation occurs postzygotically. Clearly if the imprinting of *H19* is perturbed in these circumstances, it will reveal the existence of postzygotic modifiers in imprinting of endogenous genes, and expose the relationship between imprinting of *H19* and *Igf2*. There is also interest in the fate of such imprints in embryonic stem cells of normal, androgenetic and of parthenogenetic origin. These studies would allow evaluation of the stability of imprints.

6. EMBRYONIC STEM CELLS AND THE NATURE OF PRIMARY IMPRINTS

Embryonic stem cells are derived from the primary epiblast cells which are present in blastocyts. These cells are pluripotent and possibly totipotent and derived at a time when there is a marked decline the levels of DNA methylation (Evans & Kaufman 1981; Monk 1990; Nagy *et al.* 1990; Li *et al.* 1992). Whether the loss of genome-wide methylation during preimplantation development is prerequisite for pluripotency of the epiblast cells remains to be determined. Nevertheless, the dramatic changes in this major epigenetic modification is of interest from the point of view of parental imprints. In the absence of genome wide methylation, embryonic stem (ES) cells may contain the less stable primary imprints. As discussed above, methylation of the *H19* paternal gene occurs postzygotically and only the primary germline specific imprints may be present prior to methylation. It is important to determine the nature and stability of this imprint. Embryonic stem cells are very useful in this regard.

The phenotypic and molecular properties of androgenetic, parthenogenetic and control cells can be compared after passaging the cells *in vitro* to test for the stability of the parental imprints (Mann *et al.* 1990; Mann 1992). In addition, these properties can be compared directly with those of the epiblast cells in the inner cell mass (ICM) from which the ES cells are derived. The most obvious test which has been applied to check for the properties of these cells is by assessing their effects in chimeras. Androgenetic cells derived from ICM induce marked growth enhancement and severe abnormalities of skeletal elements in chimeras (Barton *et al.* 1991; Mann 1992). This is consistent with the propensity of these cells to contribute to the many derivatives of the mesenchymal pluripotent cells such as chondrocytes and skeletal muscle. These very profound phenotypic effects are also observed when androgenetic ES cells are used to produce chimeric embryos (Mann *et al.* 1990). These phenotypic effects are observed even when these cells have undergone considerable passaging *in vitro* and with euploid as well as aneuploid cells. These remarkable similarities in properties between ES and ICM derived epiblast cells indicate that the ES cells probably retain their parental imprints, assuming that appropriate imprinting does indeed occur in androgenetic embryos in the absence of a maternal genome. Although androgenetic cells can contribute to a variety of somatic cell types in chimeric embryos, there is as yet no evidence to show if they can contribute to germ cells.

Parthenogenetic ES cells by contrast may not retain their phenotypic properties. For example, the presence of parthenogenetic cells derived from the inner cell mass in chimeras produces growth retardation (Solter 1988; Surani *et al.* 1990). Furthermore there is a remarkable lack of contribution by these cells to skeletal muscle (Surani *et al.* 1990). However, parthenogenetic ES cells do not cause such growth retardation and they contribute to skeletal muscle (Mann 1992; N. D. Allen, S. C. Barton & M. A. Surani, unpublished results). There is at present little information on precisely why this difference exists and whether the change can be attributed to alterations in the maternal imprints in ES cells. However, a major point of similarity between parthenogenetic ES cells and ICM cells is seen in their ability to contribute to the germline resulting in viable oocyte (Surani *et al.* 1990; Mann 1992).

Further studies are needed to elucidate the properties of the ES cells with regard to expression of imprinted genes. In particular it is important to know how the androgenetic cells are able to retain relatively stable phenotypic properties while the parthenogenetic cells apparently do not. We assume that the primary parental imprints are retained in the pluripotent normal ES cells. The recent derivation of stem cells directly from primordial germ cells (the so called EG cells) should provide a valuable source of material with which to explore the question of primary imprints and to compare their properties with a variety of ES cells of different genotypes (Matsui *et al.* 1992; Resnick *et al.* 1992). The normal ES cells are obtained from the epiblast cells from which both the primordial germ cells and somatic cells are derived.

7. IMPRINTING IN THE HUMAN

There is compelling evidence now to suggest that germline specific epigenetic modifications occur in the human with important consequences for some forms of genetic disorders. This is particularly well characterised for the human chromosomal regions which correspond to the mouse chromosome 7 (Cattanach & Beechey 1990; Bartolomei & Tilghman 1992). The distal end of the mouse chromosome 7 is syntenic with the human chromosome 11q13 and 11q15. The latter is particularly interesting since both *Igf2* and *H19* map to this region and is the potential contributor to the Beckwith–Wiedemann syndrome (Bartolomei & Tilghman 1992). This disorder is a foetal overgrowth syndrome associated with certain embryonal tumours such as rhabdomyosarcomas and Wilms. In sporadic cases of these diseases, paternal isodisomy of 11p15.5 is observed. The role of IGF2 in these instances is predictable since such isodisomy will result in an excess of IGF2.

The other significant human genetic disorders attributed to genetic imprinting are the Prader–Willi and Angelman syndromes which are due to parental origin specific deletions in the 15q11–13 regions and to uniparental disomies with respect to the maternal and paternal regions, respectively (Nicholls *et al.* 1992; Barr *et al.* 1992). A number of cDNA clones have now been obtained from this region which map to the mouse chromosome 7 in the vicinity of the albino locus. Indeed a new imprinted gene has now been identified in this region called SNRNP and is expressed specifically in the brain. The fact that the Prader–Willi and Angelman syndromes manifest as behavioural disorders suggests that this imprinted gene may have an important role in the genesis of Prader–Willi syndrome.

8. CONCLUSIONS

Genomic imprinting is a reversible germline-specific epigenetic modification. The identification of the first imprinted genes will allow an understanding of the mechanism of this process. In particular it is necessary to identify the nature of epigenetic modifications, their mode of inheritance, and their influence in regulating gene expression. Understanding such chromosomal controls of gene expression will provide an important insight into gene regulation in general. These studies should also broaden our understanding of the fundamental concept of totipotency as work on ES cells of different genotypes is extended. The critical steps in the reversal and re-introduction of parental imprints in the germline and their subsequent modifications after fertilization are crucial for normal development and for restoring totipotency. Aberrant germline events, for example in the process requiring erasure of previous epigenetic modifications could have detrimental consequences. These studies will ultimately demonstrate the importance of germline specific epigenetic inheritance in development and disease.

The studies described in this article were supported by the AFRC. P.A.J. was partly supported by an Eleanor Roosevelt Fellowship from the International Union Against Cancer during a sabatical in Cambridge. A.F.-S. was an AFRC Babraham Research Fellow, H.S. is supported by a Wellcome Trust Fellowship and N.D.A. by an AFRC Transgenic Animal Programme Grant. W.R. is a Fellow of the Lister Institute of Preventive Medicine.

REFERENCES

Allen, N.D., Norris, M.L. & Surani, M.A. 1990 Epigenetic control of transgene expression and imprinting by genotype-specific modifiers. *Cell* **61**, 853–861.

Allen, N.D. & Mooslehner, K.A. 1992 Imprinting transgene methylation and genotype-specific modification. *Semin. Devl Biol.* **3**, 87–98.

Barlow, D.P., Stoger, R., Herrmann, B.G., Saito, K. & Schweifer, N. 1991 The mouse insulin-like growth factor type-2 receptor is imprinting and closely linked to the time locus. *Nature, Lond.* **349**, 84–87.

Barr, J., Jones, J., Beechey, C. & Cattanach, B. 1992 Imprinting studies central region of mouse chromosome F. Mammalian Genetics & Development Workshop, London.

Bartolomei, M.S., Zemel, S. & Tilghman, S.M. 1991 Parental Imprinting of the mouse *H19* gene. *Nature, Lond.* **351**, 153–155.

Bartolomei, M.S. & Tilghman, S.M. 1992 Parental imprinting of mouse chromosome of *Semin. Devl Biol.* **3**, 107–117.

Barton, S.C., Ferguson-Smith, A.C., Fundele, R. & Surani, M.A. 1991 Influence of paternally imprinted genes on development. *Development* **113**, 679–688.

Bird, A.P. 1986 *CpG*-rich island and the function of DNA methylation. *Nature, Lond.* **321**, 209–213.

Brunkow, M.E. & Tilghman, S.M. 1991 Ectopic expression of the H19 gene in mice causes prenatal lethality. *Genes Dev.* **5**, 1092–1101.

Cattanach, B.M. & Beechey, C.V. 1990 Autosomal and X-chromosome imprinting. *Development* (Suppl.), 63–72.

Chaillet, J.R. 1992 DNA methylation and genomic imprinting in the mouse. *Semin. Devl Biol.* **3**, 99–105.

Chaillet, J.R., Vogt, T.F., Beler, D.R. & Leder, P. 1991 Parental-specific methylation of an imprinted transgene is established during gametogenesis and progressively changes during embryogenesis. *Cell* **64**, 77–84.

De Chiara, T.M., Robertson, E.J. & Efstratiadis, A. 1991 Parental imprinting of the mouse insulin-like growth factor II gene. *Cell* **64**, 849–859.

De Chiara, T.M., Efstratiadis, A. & Robertson, E.J. 1990 A growth deficiency phenotype in heterozygous mice carrying an insulin-like growth factor II gene disrupted by targetting. *Nature, Lond.* **345**, 78–80.

Engler, P., Hass, D., Puikert, C.A., Doglio, L., Glymour, M., Brinster, R. & Storb, U. 1991 A strains specific modifier on mouse chromosome 4 controls the methylation of independent transgene lon. *Cell* **65**, 939–947.

Ferguson-Smith, A.C., Cattanach, B.M., Barton, S.C., Beechey, C.V. & Surani, M.A. 1991 Embryological and molecular investigations of parental imprinting on mouse chromosome 7. *Nature, Lond.* **351**, 667–670.

Grant, S.G. & Chapman, V.M. 1988 Mechanisms of X-chromosome regulation. *A. Rev. Genet.* **22**, 199–223.

Jahner, D. & Jaenisch, R. 1984 DNA methylation in early mammalian development. In *DNA methylation* (ed. A. Razin, H. Cedar & A. D. Riggs), pp. 189–210. New York: Springer.

Kafri, T., Ariel, M., Brandeis, M., Shemer, R., Urven, L., McCarrey, J., Cedar, H. & Razin, A. 1992 Developmental pattern of gene specific DNA methylation in the mouse embryo and germ line. *Genes Dev.* **6**, 705–714.

Klose, J. & Reik, W. 1992 Expression of maternal and paternal phenotypes at the protein level. *Semin. Devl Biol.* **3**, 119–126.

Li, E., Bestor, T.H. & Jaenisch, R. 1992 Targeted mutation of the DNA Methyltransferase gene results in embryonic lathality. *Cell* **69**, 915–926.

Mann, J.R., Gadi, I., Harbison, M.L., Abbondanzo, S.J. & Stewart, C.L. 1990 Androgenetic embryonic stem cells are pluripotent and cause skeletal defects in chimeras. Implications for genomic imprinting. *Cell* **62**, 251–260.

Mann, J.R. 1992 Properties of androgenetic and parthenogenetic mouse embryonic stem cell lines; are genetic imprints conserved? *Semin. Devl Biol.* **3**, 77–85.

Matsui, Y., Zsebo, K. & Hogan, B.L.M. 1992 Derivation of pluripotent embryonic stem cells from murine primordial germ cells in culture. *Cell* **70**, 841–847.

Monk, M. 1988 Genomic imprinting. *Genes Dev.* **2**, 189–192.

Nagy, A., Gocza, E., Merentez Diaz, E., Prideaux, V.R., Ivanyi, E., Markkula, M. & Rossant, J. 1990 Embryonic stem cells alone are able to support fetal development in the mouse. *Development* **110**, 815–821.

Nicholls, R.D., Rinchik, E.M. & Driscoll, D.J. 1992 Genomic imprinting in mammalian development. Prader-Willi and Angelman syndromes as disease models. *Semin. Devl Biol.* **3**, 139–152.

Reik, W., Howlett, S.K. & Surani, M.A. 1990 Imprinting by DNA methylation: from transgenes to endogenous gene sequences. *Development* (Suppl.), 99–106.

Resnick, J.L., Bixler, L.S., Cheng, L. & Donovan, P.J. 1992 Long-term proliferation of mouse primordial germ cells in culture. *Nature, Lond.* **359**, 50–52.

Sanford, J.P., Clark, H.J., Chapman, V.M. & Rossant, J. 1987 Differences in DNA methylation during oogenesis and spermatogenesis and their persistence during early embryogenesis. *Genes Dev.* **1**, 1039–1043.

Sasaki, H., Allen, N.D. & Surani, M.A. 1992 DNA methylation and genomic imprinting in mammals. In *DNA methylation: biological significance* (ed. J. P. Jost & H. P. Saluz). Basel: Birkhauser. (In the press.)

Sasaki, H., Jones, P.A., Chaillet, J.R., Ferguson-Smith, A.C., Barton, S.C., Reik, W. & Surani, M.A. 1992 Parental imprinting: Potentially active chromatin of the repressed maternal allele of the mouse insulin-like growth factor II (*Igf2*) gene. *Genes Dev.* **6**, 1843–1856.

Searle, A.G. & Beechey, C.V. 1990 Genomic imprinting on mouse chromosome 7. *Genet. Res. Camb.* **56**, 237–244.

Solter, D. 1988 Differential imprinting and expression of maternal and paternal genomes. *A. Rev. Genet.* **22**, 127–146.

Surani, M.A. 1991 Genomic imprinting: developmental significance and molecular mechanism. *Curr. Opin. Genet. Dev.* **1**, 241–246.

Surani, M.A., Allen, N.D., Barton, S.C., Fundele, R., Howlett, S.K., Norris, M.C. & Reik, W. 1990 Developmental consequences of imprinting of parental chromosomes by DNA methylation. *Phil. Trans. R. Soc. Lond.* B **326**, 313–327.

Surani, M.A. & Reik, W. 1992 Genonic imprinting in mouse and man. *Semin. Devl Biol.* **3**, 73–75.

5

The germinal inheritance of epigenetic information in plants

R. Jorgensen

SUMMARY

Germinally transmissible variants that arise by the anomalous imposition of developmental information on the genome are not uncommon in plant genetics, although they are often ignored. Better understanding of such variants is believed to be important because they appear to reflect basic features of developmental control processes. This paper briefly reviews classical genetic studies of such variants in plants, then discusses recent work on the genetic behaviour of plant transgenes, the results of which parallel and extend the classical genetic studies of these phenomena. An attempt is made to explain the molecular basis of these phenomena in terms of modern hypotheses on the dynamic organization of chromatin.

1. INTRODUCTION

The classical view of spontaneous mutation is that it is an undirected process yielding mutant alleles that are stable and qualitatively different from their parent alleles. While this view accurately describes most genetic variants, there are also genetic variants that arise in directed fashion, i.e. by an imposition of epigenetic (developmental) information on the germ line. The term epiallele has been suggested to refer to each different form of a gene that arises in this fashion (Kermicle 1978); Holliday (1987) has suggested a similar term, epimutation, to refer to genetic variants due to DNA modifications. In this paper the definitions of these terms will be modified somewhat so that both refer to the same set of variants†.

Besides being transmissible through the germ line, epialleles are often reversible and epimutable, i.e. they change into new, distinct epialleles. Thus, the process of epimutation can be either cyclical or progressive. Epialleles that exhibit progressive change sometimes produce a continuously varying series of new epialleles which continue to be epimutable. Some epialleles are now known to be correlated with DNA modifications such as cytosine methylation, but whether this is a cause or an effect of epiallelism is still not known (see Selker (1990*a*) for discussion of this problem).

The study of epiallelism in plants has a long and distinguished history as a result of pioneering genetic studies begun in the early 1950s by Brink, McClintock, and Peterson, and followed later by Coe and Kermicle, utilizing maize kernel pigmentation genes. Important early summaries of this body of work are: Brink (1960), McClintock (1965), Peterson (1966), Coe (1966), and Kermicle (1978). By investigating the manner in which new epialleles arise, these workers discovered that three basic assumptions of classical Mendelian genetics are violable. These assumptions are: (i) that in germ line cells the hereditary substance is unaltered genetically by developmental processes; (ii) that following the passage of two alleles through the same nucleus for one heterozygous generation, each allele emerges unaltered in segregating progeny; and (iii) that gene expression following passage of a gene through a male gametophyte is equivalent to gene expression following passage of the same gene through a female gametophyte. The first assumption was found to be violated in the phenomenon of activity cycling in transposable elements, the second in the phenomenon of paramutation, and the third in the phenomenon of parental imprinting. Each of these will be briefly summarized below.

Because epimutation is a directed process, biologists have long recognized that it might reflect fundamental mechanisms of plant developmental control, perhaps for the establishment and maintenance of stable developmental states; and epiallelism has been invoked to explain processes as varied as the transitions from juvenility to adult vegetative and adult reproductive phases, the elaboration of pigmentation

† Because of the close similarity between the terms epiallele and epimutation, as well as the overlap in their definitions, it would be difficult to use both terms without modifying them to be consistent with one another. For the purposes of this paper, then, both terms will be used to refer to the same set of genetic variants, inclusive of both DNA modifications and developmentally imposed genetic variants. This is a broad definition, but the lack of precise information regarding the molecular cause of most heritable epigenetic variants, coupled with the benefits of consistent terminology makes this broad definition preferable for now. The term epimutation will be used also to refer to the process by which epialleles and epimutations arise, and epimutability is a property of any gene that is subject to the process of epimutation. Genomic imprinting is a broader term that includes germinally heritable epigenetic variants, but also refers to events that are somatically heritable, but not germinally heritable. It has been applied mostly in animal systems (Monk & Surani 1990), but also has use in plant systems (Federoff *et al.* 1989; Matzke & Matzke 1993).

patterns, and cytokinin habituation (Brink 1962; Poethig 1990; McClintock 1967*b*; Fedoroff *et al.* 1989; Meins 1989).

2. CYCLIC PHASES OF ACTIVITY IN TRANSPOSABLE ELEMENTS

McClintock discovered cycling epialleles of maize transposable elements (TEs). She referred to these as changes in phase of the TEs to distinguish them from stable, undirected mutations in TEs (which she called changes in state and have been shown to be caused by DNA sequence alterations (reviewed by Fedoroff *et al.* (1989)). Among studies of maize TEs exhibiting cyclical changes in phase (reviewed by Fedoroff (1983); see also Martiennsen *et al.* (1990)), McClintock's and Peterson's studies of the *Spm* element (also known as *En*) provide the most thorough understanding of this phenomenon.

Phases of *Spm* activity can be monitored visually and non-invasively by observing their effects on, and control over, an anthocyanin pigmentation gene into which the element (or its non-autonomous derivative) is inserted. Although different phases of activity of the *Spm* element are heritable through meiosis, they are also reversible and highly changeable. Different phases of *Spm* exist in which the element is active, inactive, or exhibiting intermediate expression. Among phases with intermediate expression, McClintock and Peterson described a very diverse array of phases that differ complexly with respect to the pattern of pigment distribution throughout the plant.

An especially significant observation is that the epimutability of these different epialleles of the *Spm* element varies from one to the next and that epimutation is not stochastic, but is dependent on developmental context. Another important feature of *Spm*'s phase change behaviour is the fact that an active *Spm* element can preset certain responsive, non-autonomous *Spm* elements for a new expression pattern (phase) which will be expressed at a later time in development, even in the absence of the element that causes the presetting. This setting typically is erased during sexual reproduction.

With regard to the mechanism behind the cycling phenomenon, McClintock (1967*a*) proposed that the phase of an element has both a consequence and a cause relating to normal mechanisms controlling gene expression. The proposed consequence of a particular phase is that it may affect the positional (or functional) domains in which the element itself will be located in the nucleus of particular cell types, resulting in developmentally varying expression patterns. Reciprocally, certain nuclear domains may cause a change in phase through the imposition of developmental information on the element. Thus, as McClintock (1967*a*) suggested 'a relatively simple sequence of events' involving reciprocal influences between genes and domain-specific nuclear features might lead to a complex series of expression patterns. Technologies for determining the locations of genes in the nuclei of different cells in order to detect such repositioning of genes are becoming available in some systems (e.g. Manuelidis 1990), offering the hope that this part of McClintock's suggestion for *Spm*'s phases can be tested in the near future.

The molecular basis for epimutations in several maize TEs, including the *Spm* element, has been explored by monitoring the methylation of cytosines with methylation-sensitive restriction enzymes. The first published demonstration of a correlation between methylation and inactivation of a TE was by Bennetzen (1985), and similar correlations have been found in several other systems since (summarized by Dennis & Brettell 1990; Martienssen *et al.* 1990). In the case of *Spm*, the pattern of methylated cytosines appears to vary in concert with changes in the element's phase of activity, both developmentally and germinally (Banks & Fedoroff 1989), suggesting that methylation pattern is a result, or maybe a component, of the cell-type-specific domains of the nucleus predicted by McClintock.

3. PARAMUTATION

The phenomenon of paramutation is as remarkable as is the cycling and presetting of activity phases of transposable elements. Paramutation results from the action of particular alleles possessing paramutagenic properties on certain other alleles with paramutable properties. The outcome is the alteration of the paramutable allele into a new paramutant allele whose phenotypic expression is germinally heritable, and is expressed even in the absence of the paramutagenic allele. In other words, paramutation is an interallelic interaction resulting in the epimutation of one allele as a result of the presence of the other. Although paramutation seems to have been observed in plants almost since the beginning of genetics, it was generally overlooked until R.A. Brink (1960) carefully reconsidered and defined, and then publicized the phenomenon. Paramutation is often considered to be enigmatic because it is an apparent violation of one of the basic tenets of genetics: that allelic genes, though they may affect one another's expression when present in the same nucleus, emerge unaltered upon segregation in sexual progeny. (For a recent review of paramutation, see Dooner *et al.* (1991).)

Paramutation at the maize *R* locus exhibits unusual patterns of instability. Paramutants of *R* appear progressively in development and change progressively and heritably over successive sexual generations, either decreasing in expression in the presence of a paramutagenic allele, or increasing (gradually returning to maximal, stable expression) in its absence. Many *R* paramutants become relatively stable in homozygous condition. Thus, like epimutation, paramutation of *R* exhibits reversibility and heritability, and is also progressive and quantitative. On the basis of these unusual properties, Brink speculated that the mechanism behind paramutation would involve changes not in the DNA, but in the organization of the chromatin encasing the affected genes. This led him to propose the concept of a paragenetic function for chromatin, as well as the term parachromatin for the type of chromatin fulfilling that function

(Brink 1960). The DNA methylation state of genes involved in paramutation has not yet been reported in the literature, although unpublished data indicate a correlation between methylation and paramutation at the *R* locus (M. Alleman & J. L. Kermicle, personal communication).

The presetting of phases of *Spm* somewhat resembles paramutation. Presetting alters the element with respect to its expression at a later time in development and does not require the presence of the other element at the time of expression. However, unlike paramutation, the setting usually is erased during sexual reproduction. Exceptions in which erasure of the setting do not occur are analogous to paramutation with the paramutagenic agent being an active *Spm* element, which can be either allelic or non-allelic to the 'paramutable' preset element. McClintock's interest focused on the erasure process because it indicated that presetting was subject to a developmental rule that also applied to any epigenetic process in normal development, i.e. reversal during the life cycle of the plant. In contrast, Brink's interest was focussed on the heritability of epigenetic changes.

4. PARENTAL IMPRINTING

Parental imprinting was first described in animals for the differential silencing or elimination of maternal and paternal chromosomes. Through further investigation of paramutant *R* alleles in maize, Kermicle discovered the first example in eucaryotes of parental imprinting of a single gene (Kermicle 1978). This effect was limited to the endosperm tissue of the maize kernel, the triploid product of a fertilization event separate from that producing the embryo. Thus, no grandparental imprinting effects are observed. The lack of imprinting in embryonic tissues (which are the result of a separate fertilization event in maize) indicates the precise control over imprinting that is exerted by the organism (Kermicle & Alleman 1990). Genetic variation for the control of imprinting also exists in maize at an unlinked modifier locus.

5. TRANSGENE EPIMUTATION

Transgenes offer an additional opportunity to study epimutation phenomena in plants. The ability to vary the chromosomal location and sequence context of a gene could have a significant effect on our understanding of, and control over the factors involved. Although it is not yet possible to manipulate sequences in the plant genome at will (homologous recombination technologies that have practical application have yet to be developed in plant systems), there is already considerable information available on epimutation of transgenes.

The clearest demonstration of transgene epimutation in plants is provided by the use of genes for anthocyanin pigmentation. Because maize transformation–regeneration systems have been developed only recently, the available information on pigment transgenes has been obtained mainly in petunia flowers. The best example involves a maize anthocyanin gene that was used to complement a pigmentation defect in an anthocyaninless petunia line (Meyer *et al.* 1992). It was found that the expression of the (single copy) transgene, though initially judged to be stable in the primary transformant and in a homozygous line derived from it, was subject to various types of epimutation. Thirty thousand hemizygous progeny of a single homozygous plant were scored for flower colour, 11 000 of which were monitored throughout an entire growing season, and the following observations were reported.

(1) A single classical mutation was uncovered, a small deletion within the coding sequence of the transgene.

(2) Plants producing variegated (red sectors arising on a white background) flowers appeared at a frequency of 2×10^{-3}. None of these resulted from the insertion of transposable elements, or from any other DNA rearrangement. Rather, all variegated plants appeared to be the result of epimutation because each showed hypermethylation of the promoter. Red sectors were inferred to be the result of somatic reversion of the white, epimutant phenotype. Further genetic analysis of these plants is required to determine the heritability and stability of these phenotypes.

(3) Most plants in the population showed weak or no transgene expression, and this correlated with methylation of the transgene promoter.

(4) 0.3% of those plants monitored throughout the growing season exhibited somatic loss of expression of the transgene (white sectors on a red background), again correlated with promoter methylation.

(5) The phenotypic expression and promoter methylation of the transgene were both subject to environmental and developmental influences. Interestingly, it could be shown that the age of the parent plant determined the susceptibility of transgene expression to environmental influences in the next generation, indicating that heritable alterations tend to occur late in development. Parallels with the epimutation behaviour of maize transposable elements are obvious. Clearly, then, pigment transgenes can be useful for investigating epimutations in plants and broadening the base of information on this process.

The fact that a heterologous pigment transgene exhibits a high degree of epimutability, while the endogenous petunia pigment genes generally do not, suggests that it is the sequence context of the transgene that determines its epimutability, as is proposed for animal transgenes. In maize, transposable element insertions into anthocyanin genes were the basis of epimutability, which in any case was limited to the TE. On the other hand, paramutation at the *R* and *B* loci in maize does not involve TE insertions, as far as is known, but might depend on particular sequence organization patterns. Structural analysis of different alleles of the *R* and *B* loci might illuminate this question, and the analysis of *R* and *B* transgenes in maize also might be very informative.

6. INTERLOCUS AND INTERALLELIC TRANSGENE INTERACTIONS

Matzke *et al.* (1989) discovered interactions between

unlinked transgene loci which shared some stretches of homology with each other, resulting in one locus being inactivated and methylated in the presence of the other. Both the inactive state and the associated methylation of the first locus are reversible in the absence of the second locus, but this reversion is gradual, occurring over several generations (Matzke & Matzke 1990, 1991). This paramutation-like interaction between unlinked transgenes has been observed with another pair of transgenes by Goring *et al.* (1991), although no methylation was detected in that case. Paramutation resulting from an interaction between alleles of a transgene locus differing in sequence organization has been demonstrated by J. English and J. Jones (personal communication). Also, epialleles of a transgene locus can interact to result in the apparent paramutation of one epiallele by another (R. Jorgensen, unpublished data). These observations of paramutation in transgene–transgene interactions would appear to rule out the possibility that a specialized protein encoded by the paramutagenic locus (perhaps analogous to the SD product involved in meiotic drive in *Drosophila*) of the paramutagenic gene is responsible for paramutation, as might be possible for endogenous paramutagenic genes.

Interlocus effects are not limited to transgene–transgene interactions, as there are now many examples whereby a transgene can cause the suppression of a homologous, endogenous plant gene (see, for example, Napoli *et al.* (1990); van der Krol *et al.* (1990); Smith *et al.* (1990)). An intriguing feature of transgene–endogenous gene interactions is that commonly the expression of both genes is suppressed (referred to as co-suppression), even though the genes are driven by unrelated promoters. However, it is not known whether transgenes suppress homologous, endogenous genes via the same mechanism as transgenes suppress other, homologous transgenes. In fact, no paramutation-like suppression of an endogenous gene after segregation of the homologous transgene has been reported. The available data on *trans*-inactivation by transgenes, along with several hypotheses attempting to explain them, have been reviewed recently (Jorgensen 1992).

7. EPIALLELIC VARIATION IN ORGANIZED FLOWER PIGMENTATION PATTERNS IN TRANSGENIC PETUNIAS

Interference with the expression of an endogenous plant gene by a homologous transgene can be surprisingly complex. This has been illustrated best with flower colour genes, especially the chalcone synthase gene (Napoli *et al.* 1990; van der Krol *et al.* 1990; R. Jorgensen, unpublished data). Chalcone synthase is involved in the biosynthesis of anthocyanin pigments, and co-suppression of an endogenous chalcone synthase gene and the transgene results in the production of white flowers or flowers with developmentally determined colour patterns. The flower pigmentation patterns are comprised of white sectors which are caused by the co-suppression of the expression of the transgene and the endogenous chalcone synthase gene. These patterns are highly organized and appear to be determined by cells located either at the edges of petals or in or near the vasculature of petals, and

Figure 1. Suppression of anthocyanin pigmentation by chalcone synthase transgene. Five petunia flowers representative of non-clonal patterns produced by derivatives of a single transgenic plant are shown. For contrast, flower in upper right illustrates the shape of typical clonal sectors as might be produced through transposable element mutations. Among the five non-clonal patterns, at least three genetically distinct derivatives of a single transformant are represented by the flowers at the upper left, lower left, and lower right. The lower middle flower is genetically similar to the one at the lower right. The upper middle flower is genetically similar to the upper left.

possibly by cells underlying the epidermal cells which elaborate the pigmentation pattern (figure 1). The white sectors of each pattern are non-clonal and have diffuse boundaries, that is, the sectors are many cells wide. Furthermore, pattern boundaries are coincident in the upper and lower epidermis. Thus, elaboration of these patterns must require intercellular communication. Moreover, patterns appear to be determined similarly to (and perhaps based on) morphological patterns that determine petal shape.

What is interesting in the context of this paper is that a complex series of somatic and germinal variants appear spontaneously in these transgenic petunias. Such variants display new patterns different in form from the parental pattern. Examples are shown in figure 1. The frequency of variants is especially high following adventitious shoot regeneration from tissue explants (a procedure capable of altering methylation patterns). Many of these variants are germinally heritable, and they are reversible somatically as well as germinally, i.e. they are epialleles. DNA rearrangements have not been ruled out yet; nor is it clear whether these epialleles map to the transgene locus or the two endogenous CHS loci (CHS A & J) or another locus.

Perhaps most significant is the observation that the occurrence of epimutations can be non-random, with bursts of independent epimutations arising at the same time in mature plants. Thus, not only are these epimutations reversible, heritable and changeable, but they occur in response to developmental circumstances. Some somatically occurring epialleles are not inherited by all progeny, but are 'reset' to the original epiallelic state during sexual reproduction or in early development. In sum, this transgenic system displays the same features that McClintock described for TE activity cycling. It also illustrates the complexity and lability of gene expression patterns that can be produced by transgenes.

8. THE PARAGENETIC FUNCTION OF THE GENOME

The remainder of this paper will discuss various concepts that might be useful in understanding the process of epiallelism in plants. Earlier I described McClintock's concept of reciprocal influences between genes and domain-specific nuclear features. A related and somewhat more detailed concept came from R. A. Brink (1960, 1962) who proposed that the genome serves a paragenetic function in addition to its genetic one, the former being concerned with the development of the individual, the latter with inheritance. In Brink's words, the chromosomes not only 'carry an unchanging complement of genes in somatic cells, but also possess other, more labile components by which gene action is regulated during development'. He further suggested that a particular paragenetic state is self-maintaining in mitosis, but subject to influences from the developmental and physiological environment within the organism, i.e. it is developmentally paramutable. Thus, 'the individuality of a somatic cell is implicit in the paragenetic state of its chromosomes'.

Brink proposed the term parachromatin to refer to whatever chromosomal features are responsible for the paragenetic state. It is important to recognize that the concept of parachromatin was not meant to be limited to genes which are germinally paramutable. Rather, Brink regarded germinally transmissible paramutants only as aberrant manifestations of a fundamental process that might be relevant to the behaviour of all plant genes, and perhaps all higher eucaryotic genomes.

9. GENOMIC IMPRINTING AND CHROMATIN ARCHITECTURE

Brink's concept of parachromatin anticipated to a considerable extent the modern concept of genomic imprinting. Genomic imprinting is a term introduced by animal biologists to refer to the clonal inheritance of developmentally imposed states, as is apparent in parental imprinting and X-chromosome inactivation in mammals, and in position effect variegation and polycomb gene behaviour in *Drosophila* (for examples, see reviews by Surani (1991); Monk (1990); Henikoff (1990); Paro (1990)). Among these phenomena only parental imprinting shows any germinal inheritance, and parental imprints are generally erased each generation to be replaced by a new imprint. How this concept is applicable to epiallelism phenomena in plants has been discussed in some detail by Matzke & Matzke (1993).

Since 1960 when Brink proposed the parachromatin concept, much has been learned about the components of chromatin from both genetic and molecular studies of genomic imprinting phenomena. These studies have led to the suggestion that it is 'the architecture of chromatin, in contrast to the regulatory networks of diffusible factors', that serves 'as a cellular memory transmitting the determined state of a cell to its clonal descendants' (Paro 1990). Investigations of the complexity, dynamics and stability of multi-component DNA-protein complexes (i.e. chromatin) at specific loci are now possible for some well-characterized genes. These biochemical studies now suggest at least two types of mechanisms by which chromatin structure may be altered developmentally: dynamic competition and pre-emptive competition, which differ primarily in the stability (or resistance) of chromatin to potential disruption by newly synthesized factors (Felsenfeld 1992). Similarly, genetic and molecular studies of *Drosophila polycomb* genes and *suppressor of position effect variegation* genes indicate that clonally propagated states involve heterochromatin proteins (Paro 1990). Thus, chromatin architecture could be the basis for a stable, yet flexible, ontogenetic cell memory.

Recently, greater resolution of chromatin organization has been achieved by the discovery and characterization of a class of chromatin components known as 'global activators'. These have been best characterized in yeast but found to be widespread and highly conserved among eucaryotes. These global factors interact with both gene-specific transcription factors and basic chromatin components such as histones and

HMG proteins. One of the functions of global activators is a helicase activity. It has been suggested that the primary role of the helicase function is to reprogram chromatin and to establish both repressed and active states of chromatin (Travers 1992; Winston & Carlson 1992).

10. THE ORIGIN OF EPIALLELES AND GENE IMPRINTS

What then might be the origin of epigenetic variants that have sufficient stability as to persist through germinal transmission? From the perspective outlined above, it is conceivable that certain multimeric chromatin structures could be perturbed through developmental fluctuations in the levels or modifications of transcription factors or other chromatin components. These perturbations might lead in turn to alterations in DNA methylation patterns due to exposure or protection of DNA. DNA methylation provides an attractive mechanism for the inheritance of paragenetic information because DNA methylation is known to be heritable in some cases. Mechanistically this heritability can be explained by the fact that replication of DNA produces hemi-methylated DNA which is then recognized by a maintenance methylase to preserve the methylation pattern (Holliday 1987). Assuming that DNA methylation patterns are sufficient to maintain local chromatin organization patterns through multiple generations of cell division, it is conceivable that methylation could stabilize a developmentally instigated perturbation of chromosome organization, and also that interference with the maintenance methylation process is one possible cause of epimutations.

It is also possible that DNA methylation is not always sufficient, or even necessary for heritability of epimutations, and that other mechanisms, such as higher order levels of chromosome organization, might be responsible for the inheritance of 'paragenetic' information (Riggs & Pfeiffer 1992). Thus, it may also be necessary to look beyond the local aspects of chromatin organization in order to understand epiallelism. The types of chromosomal domains that might be responsible for the paragenetic state of a gene could be functional, spatial or temporal. Functional differences between alternative paragenetic states (i.e. epialleles) of a given gene could involve DNA methylation as well as the presence, the concentration and the modification state of certain chromatin components. Spatially distinct alternative states could arise via movement of a chromosome segment between different nuclear locations capable of conferring different gene expression potentials. Temporal differentiation of 'parachromatin' could be achieved by modifying the time of DNA replication in the cell cycle, late replication being commonly associated with lack of transcription, as has been discussed by Riggs & Pfeiffer (1992).

It is important to address the question of why some genes are epimutable and most are not. Assuming that the stability and flexibility (i.e. epimutability) of the alternative chromatin structures at a locus depends to some degree on the DNA sequence context of and around the gene, it would be expected that the sequence context of most genes would have been selected by evolution to resist any imposition of germinally heritable alterations during ontogeny. Occasionally new alleles will arise that fail to resist such forces, i.e. alleles that are epimutable or paramutable. Although these may not survive in evolutionary time, they would be observed as unstable variants by plant breeders and geneticists. The fact that the most closely studied examples of epiallelism in plants involve kernel and plant pigmentation in maize may not be coincidental. The well-known cultural importance of pigmented maize lines to the indigenous peoples of the Americas may be responsible for much of the genetic diversity behind this phenotype that is available to modern geneticists. In the case of plant transgene loci, epimutability, paramutability, and paramutagenicity might be unusually common due to the abnormal sequence context in which they often may lie.

(Note: As was indicated in the footnote on the breadth of the definition of the terms epiallele and epimutation, DNA rearrangements are not excluded as possible causes of these phenomena. DNA rearrangements that could cause epimutations of this broad definition would have to be of a type that are reversible or highly unstable. An example would be phase variation in *Salmonella typhimurium* or mating type determination in *Saccharomyces cerevisiae*. In the case of TEs, these elements are capable of mediating transposition, inversion, and deletion events that can generate a diverse series of new alleles which are typically stable and irreversible. Events such as these are known to be responsible for changes in state, i.e. classical mutations of the TE, as well as mutations of adjacent genes (Federoff 1983). McClintock distinguished changes in phase from changes in state by the cyclical behaviour of the former. It is now known that changes in phase of TEs do not involve DNA sequence alterations, but are correlated with DNA methylation (Banks & Fedoroff 1989). Nevertheless, it is important to note the possibility that reversible sequence rearrangements could produce the same phenomenology as changes in phase and meet Kermicle's definition of epiallele. In fact, the occurrence of such rearrangements could be responsive to developmental or environmental influences because the expression of transposase can be developmentally controlled. Thus, it is essential that the investigation of the mechanisms behind epiallelism test for the involvement of DNA rearrangements.)

11. MECHANISMS BEHIND INTERALLELIC AND INTERLOCUS INTERACTIONS

(*a*) *Chromatin perturbation and competition for diffusible factors*

It is conceivable that one gene (or allele) could affect another through competition for diffusible factors. For example, if two genes differ in their ability to compete for such factors, one might perturb the other's chromatin organization. Or, in the case that a gene's interaction with a factor is dependent on a certain threshold level of that factor, introduction of additional gene copies could reduce the amount of the factor that is available to each gene copy below the needed threshold level. Because most transgene loci do not exhibit an interlocus *trans*-interaction, Matzke & Matzke (1990) suggest that any such diffusible factors may be diffusion-limited such that the relative nuclear locations of two genes would determine whether any *trans*-interaction could occur. A difficulty with this model is that it does not explain the specificity, i.e. why no other genes appear to paramutated.

(*b*) ***Homology-based interactions between nucleic acids***

Recent discoveries in fungal and animal genetics suggest two additional hypotheses for how paramutation might occur. Both hypotheses propose that nucleic acids might interact with the paramutable gene in a homology-dependent manner, and that the result of this interaction could be a perturbation of chromatin organization or DNA methylation pattern. These hypotheses and their implications have been described in a recent review (Jorgensen 1992) and so they will be summarized here only briefly.

The first possibility is that transcripts from a paramutagenic gene might form an RNA–DNA hybrid (perhaps triple-stranded) with the paramutable gene (also considered by van der Krol *et al.* (1990) for transgene-caused gene silencing). Such an RNA might be either a sense or an antisense strand. It is at least conceivable that RNA base-pairing with a gene *in trans* could perturb chromatin structure and so cause its paramutation. Such a mechanism remains to be demonstrated in plants and remains purely speculative for now. So-called 'anti-gene' RNAs are the subject of serious attempts to attenuate animal gene expression (reviewed by Helene (1991)).

The second possibility for nucleic acid interactions is that homologous genes might pair directly with one another. Such pairing might be either allelic or ectopic (non-allelic). Very efficient ectopic pairing is known to occur in the RIP system in *Neurospora*, as well as in ectopic gene conversion in yeast and *Drosophila*. The filamentous fungus *Neurospora crassa* exhibits a phenomenon resembling the transgene-caused gene silencing phenomenology of plants that is called repeat-induced point mutation (RIP) whereby any duplicated sequence becomes mutated and methylated at a premeiotic stage of its reproductive phase. This process appears to require pairing of homologous sequences followed by recognition of the pair by a system for DNA and chromatin modification (Selker 1990*b*). Analogies have been drawn between RIP and ectopic gene conversion in yeast (Haber *et al.* 1991). Unlike transgene-caused gene silencing in plants, RIP occurs only rarely, if at all, in somatic diploid fungal cells. Evidence of a 'slow RIPping' process in animals comes from the comparison of rates of C-to-T mutations in multi-copy genes as compared to single-copy genes. Considering that these mutations must have occurred in the germ-line, Drake (1991) has suggested that both 'slow ripping' in animals and 'fast ripping' in fungi are limited to the haploid phase of the life cycle.

For paramutation to occur as a result of any nucleic acid interactions, chromatin organization and DNA methylation properties of the target gene must be susceptible to perturbation as a result of these interactions. Thus, whether a persistent alteration in chromatin organization or DNA methylation pattern might occur would depend on the susceptibility of the target gene, as well as on the paramutagenic features of the trans-acting RNA or DNA molecule. That most genes fail to exhibit paramutation in plants is presumed to be a reflection of the fact that natural selection generally would eliminate paramutable alleles, i.e. those whose chromatin organization is not reset during sexual reproduction. This is consistent with Brink's view of paramutation as an aberrant process reflecting the existence of certain normal features of chromatin organization or dynamics.

(*c*) ***Plants contrasted with animals***

It is intriguing that transgene-caused silencing of homologous genes has not yet been described for transgenic animals. Unless this reflects the possibility that such effects have gone unnoticed in animals, it would seem that plants possess features of gene regulation or nuclear organization. On the other hand, Monk (1990) and Tartof & Henikoff (1991) have speculated that animals may have 'cross-talk' or 'trans-sensing' mechanisms operating in somatic cells for communication between homologues with regard to the epigenetic state and history of genes. If so, then perhaps the main difference between plants and animals in this regard is only a matter of degree.

It is interesting also that no true parallel to paramutation has been revealed by animal genetics. Brink (1960, 1973) cites several possible examples of paramutation in *Drosophila* (meiotic drive and unequal crossing-over within arrays of tandem repeats); however, these would now appear to be unrelated mechanistically to paramutation. This apparent difference between plants and animals might be explained if animals restrict both the creation and the erasure of epialleles to specific developmental stages. In animals most epialleles arise through parental imprinting, i.e. during sexual reproduction (see Surani *et al.*, this symposium). The highly efficient parental imprinting in animals ensures the erasure of epialleles arising in the grandparental generation. Other than the gametic imprinting of endosperm genes, the closest parallel in plants is the efficient erasure of presetting.

Without denying the broad similarities in gene regulatory mechanisms and chromatin components known to be shared by plants and animals, it is nonetheless reasonable to ask whether their nuclei might differ significantly with respect to nuclear or chromosome organization patterns, especially in terms of higher order structures, about which little is yet known. Certainly, it is true that these phyla have been evolutionarily distinct for a very long time, having diverged prior to each kingdom's independent evolution of multicellularity and of a prolonged, specialized diploid phase in their life cycles. Thus, these phyla have had ample opportunity to evolve significant specializations in the dynamic organization of chromatin and nuclei.

12. IMPLICATIONS FOR THE ORIGIN OF DIRECTED MUTATIONS

The occurrence of selection-induced mutation processes yielding irreversible alterations in DNA sequence has been reported repeatedly in bacteria and recently in yeast (Hall 1992). The directed nature of the process of epimutation is intriguing in this regard. With regard to the origin of selection-induced mutations, one could speculate that environmental factors could induce epiallelic variants in normal genes, and that such variants, even if unstable, might predispose the gene to be either more or less sensitive to mutagenic molecular machinery, altering the rate of mutation at the locus. If a higher rate of mutation occurs, a DNA sequence variant might arise that stabilizes the new epiallelic state of the locus. It is conceivable that only those DNA alterations which in fact stabilize the epiallele would be permitted or favored.

A specific proposal of this sort has been proposed by Flavell (1991). It is based on rRNA genes in plants and the knowledge that: (i) not all rRNA genes with the same function are in the same physical state in the nucleus; (ii) these states differ with respect to their accessibility to transcription machinery; and (iii) changes in the number of tandem copies of rRNA genes are subject to phenotypic selection. Thus, it is necessary to postulate only that recombination enzymes are differentially able to produce genetic variation in different rRNA gene loci. Though less information is currently available on the nature of epialleles than on the nature of states of rRNA gene loci, the possibility that epiallelism could be related to the selection-induced mutation phenomenon is at least worth considering because the implications for evolutionary biology could be significant (see, in addition, Monk (1990)).

13. A CONCLUDING PERSPECTIVE

After consideration of the various exceptions to Mendelian principles reviewed here, it is difficult to escape the conclusion that science is still extremely ignorant about the functioning and organization of the nucleus. In spite of the remarkable advances in molecular dissection of many nuclear processes, we still can not say we really understand gene expression and control. And in spite of the powerful appeal of the principles of population genetics for explaining the prime forces in the evolution of species, the parallel phenomena of directed epimutation and selection-induced mutation suggest that perhaps we understand far less about these forces than most scientists had recently assumed. Thus, nearly a century after the rediscovery of Mendel's laws, the challenge to understand the basis for the remarkable diversity of known genetic phenomena in terms of cellular organization and function remains nearly as great as ever.

I thank Jerry Kermicle, Hugo Dooner, Virginia Walbot, Marge and Tony Matzke, and Dick Flavell for helpful discussions and comments on the manuscript. Any errors of interpretation are my own responsibility. I also thank Mary Alleman, Jerry Kermicle, Jim English, and Jonathan Jones for permission to mention unpublished information. I am grateful to the DNA Plant Technology Corporation for providing the transgenic petunia plants for my research and to the USDA for supporting part of my research for the past year.

REFERENCES

Banks, J.A. & Fedoroff, N. 1989 Patterns of developmental and heritable change in methylation of the *Suppressor-mutator* transposable element. *Devl Genet.* **10**, 425–437.

Bennetzen, J.L. 1985 The regulation of *Mutator* function and *Mu1* transposition. *UCLA Symp. Molec. Cell. Biol.* **35**, 343–354.

Brink, R.A. 1960 Paramutation and chromosome organization. *Q. Rev. Biol.* **35**, 120–137.

Brink, R.A. 1962 Phase change in higher plants and somatic cell heredity. *Q. Rev. Biol.* **37**, 1–22.

Brink, R.A. 1973 Paramutation. *A. Rev. Genet.* **7**, 129–152.

Coe, E.H. Jr 1966 The properties, origin, and mechanism of conversion-type inheritance at the *B* locus in maize. *Genetics* **53**, 1035–1063.

Dennis, E.S. & Brettell, R.I.S. 1990 DNA methylation of maize transposable elements is correlated with activity. *Phil. Trans. R. Soc. Lond.* **B326**, 217–229.

Dooner, H.K., Robbins, T.P. & Jorgensen, R.A. 1991 Genetic and developmental control of anthocyanin biosynthesis. *A. Rev. Genet.* **25**, 173–199.

Drake, J.W. 1991 Spontaneous mutation. *A. Rev. Genet.* **25**, 125–146.

Fedoroff, N.V. 1983 Controlling elements in maize. In *Mobile genetic elements* (ed. J. A. Shapiro), pp. 1–63. New York: Academic Press.

Fedoroff, N., Masson, P. & Banks, J.A. 1989 Mutations, epimutations, and the developmental programming of the maize *Suppressor-mutator* transposable element. *BioEssays* **10**, 139–144.

Felsenfeld, G. 1992 Chromatin as an essential part of the transcriptional mechanism. *Nature, Lond.* **355**, 219–224.

Flavell, R.B. 1991 Reference points in evolution: molecular processes in the genome and gene evolution; a hypothesis that the most effective members of a multigene family may be preferential substrates for mutational processes which affect gene frequencies. In *Reference points in evolution*, pp. 15–16. Madrid: Fundacion Juan March.

Goring, D.R., Thomson, L. & Rothstein, S.J. 1991 Transformation of a partial nopaline synthase gene into tobacco suppresses the expression of a resident wild-type gene. *Proc. natn. Acad. Sci. U.S.A.* **88**, 1770–1774.

Haber, J.E., Leung, W.-Y., Borts, R.H. & Lichten, M. 1991 The frequency of meiotic recombination in yeast is independent of the number and position of homologous donor sequences: implications for chromosome pairing. *Proc. natn. Acad. Sci. U.S.A.* **88**, 1120–1124.

Hall, B.G. 1992 Selection induced mutations occur in yeast. *Proc. natn. Acad. Sci. U.S.A.* **89**, 4300–4303.

Helene, C. 1991 The anti-gene strategy: control of gene

expression by triplex-forming-oligonucleotides. *Anti-Cancer Drug Desn* **6**, 569–584.

Hennikoff, S. 1990 Position-effect variegation after 60 years. *Trends Genet.* **6**, 422–426.

Holliday, R. 1987 The inheritance of epigenetic defects. *Science, Wash.* **238**, 163–170.

Jorgensen, R. 1992 Silencing of plant genes by homologous transgenes. *Agbiotech News Inform.* **4**, 265N–273N.

Kermicle, J.L. 1978 Imprinting of gene action in maize endosperm. In *Maize breeding and genetics* (ed. D. B. Walden), pp. 357–371. John Wiley and Sons.

Kermicle, J.L. & Alleman, M. 1990 Gametic imprinting in maize in relation to the angiosperm life cycle. *Development* (Suppl.), 9–14.

Manuelidis, L. 1990 A view of interphase chromosomes. *Science, Wash.* **250**, 1533–1540.

Martienssen, R., Barkan, A., Taylor, W.C. & Freeling, M. 1990 Somatically heritable switches in the DNA modification of *Mu* transposable elements monitored with a suppressible mutant in maize. *Genes Dev.* **4**, 331–343.

Matzke, M.A. & Matzke, A.J.M. 1990 Gene interactions and epigenetic variation in transgenic plants. *Devl Genet.* **11**, 214–223.

Matzke, M.A. & Matzke, A.J.M. 1991 Differential inactivation and methylation of a transgene in plants by two suppressor loci containing homologous sequences. *Pl. molec. Biol.* **16**, 821–830.

Matzke, M.A. & Matzke, A.J.M. 1993 Genomic imprinting in plants: parental effects and *trans*-inactivation phenomena. *A. Rev. Pl. Physiol. Pl. molec. Biol.* **44**. (In the press.)

Matzke, M.A., Primig, M., Tronovsky, J. & Matzke, A.J.M. 1989 Reversible methylation and inactivation of marker genes in sequentially transformed tobacco plants. *EMBO J.* **8**, 643–649.

McClintock, B. 1965 The control of gene action in maize. *Brookhaven Symp. Biol.* **18**, 162–184.

McClintock, B. 1967*a* Regulation of pattern of gene expression by controlling elements in maize. *Carnegie Instn Yb* **65**, 568–578.

McClintock, B. 1967*b* Genetic systems regulating gene expression during development. *Devl Biol. Suppl.* **1**, 84–112.

Meins, F. Jr 1989 Habituation: heritable variation in the requirement of cultured plant cells for hormones. *A. Rev. Genet.* **23**, 395–408.

Meyer, P., Linn, F., Heidmann, I., Meyer z.A., H., Niedenhof, I. & Saedler, H. 1992 Endogenous and environmental factors influence 35S promoter methylation of a maize A1 gene construct in transgenic petunia and its colour phenotype. *Molec. Gen. Genet.* **231**, 345–352.

Monk, M. 1990 Variation in epigenetic inheritance. *Trends Genet.* **6**, 110–114.

Monk, M. & Surani, M.A. (eds) 1990 *Genomic imprinting.* (*Development* 1990 Suppl.). The Company of Biologists.

Napoli, C., Lemieux, C. & Jorgensen, R. 1990 Introduction of a chimeric chalcone synthase gene into petunia results in reversible co-suppression of homologous genes *in trans. Pl. Cell* **2**, 279–289.

Paro, R. 1990 Imprinting a determined state into the chromatin of *Drosophila. Trends Genet.* **6**, 416–421.

Peterson, P.A. 1966 Phase variation of regulatory elements in maize. *Genetics* **54**, 249–266.

Poethig, R.S. 1990 Phase change and the regulation of shoot morphogenesis in plants. *Science, Wash.* **250**, 923–930.

Riggs, A.D. & Pfeiffer, G.P. 1992 X-chromosome inactivation and cell memory. *Trends Genet.* **8**, 169–174.

Selker, E.U. 1990*a* DNA methylation and chromatin structure: a view from below. *Trends Biochem. Sci.* **15**, 103–107.

Selker, E.U. 1990*b* Premeiotic instability of repeated sequences in *Neurospora crassa. A. Rev. Genet.* **24**, 579–613.

Smith, C.J.S., Watson, C.F., Bird, C.R., Ray, J., Schuch, W. & Grierson, D. 1990 Expression of a truncated tomato polygalacturonase gene inhibits expression of the endogenous gene in transgenic plants. *Molec. Gen. Genet.* **224**, 477–481.

Surani, M.A. 1991 Genomic imprinting: developmental significance and molecular mechanism. *Curr. Opin. Genet. Dev.* **1**, 141–246.

Tartof, K.D. & Henikoff, S. 1991 Trans-sensing effects from Drosophila to humans. *Cell* **65**, 201–203.

Travers, A.A. 1992 The reprogramming of transcriptional competence. *Cell* **69**, 573–575.

van der Krol, A.R., Mur, L.A., Beld, M., Mol, J.N.M. & Stuitje, A.R. 1990 Flavonoid genes in petunia: addition of a limited number of gene copies may lead to a suppression of gene expression. *Pl. Cell* **2**, 291–299.

Winston, F. & Carlson, M. 1992 Yeast SNF/SWI transcriptional activators and the SPT/SIN chromatin connection. *Trends Genet.* **8**, 387–391.

6

The regulation of human globin gene switching

F. Grosveld, M. Antoniou, M. Berry, E. de Boer, N. Dillon, J. Ellis, P. Fraser, O. Hanscombe, J. Hurst, A. Imam, M. Lindenbaum, S. Philipsen, S. Pruzina, J. Strouboulis, S. Raguz-Bolognesi and D. Talbot

SUMMARY

This paper describes the mechanism of regulation of the human β-globin on the basis of a number of natural mutations and experiments in transgenic mice. From these data we conclude that this multigene locus is regulated at a number of different levels involving specific interactions between the Locus Control Region (LCR) and the individual genes. Most important is the action of stage specific transcription factors acting on sequences immediately flanking the genes. In addition, specificity is obtained through specific interaction of the genes with the LCR and through competition of the genes for interaction with the LCR.

1. INTRODUCTION

The human β globin gene cluster spans a region of 70 kb on chromosome 11 and contains five developmentally regulated genes in the order 5′ ε,$\gamma_G\gamma_A$,δ,β 3′. In the early stages of human development, the embryonic yolk sac is the haematopoietic tissue and expresses the ε globin gene. This is followed by a switch to the γ globin genes in the foetal liver and the δ and β genes in adult bone marrow (for review, see Collins & Weissman (1984)). At all stages the output of these genes is balanced by that of the α gene locus. The important difference between the α globin genes and the β globin genes is the fact that they undergo a single switch and do not have an additional switch around the time of birth. At all developmental stages these genes are expressed at exceptionally high levels giving rise to 90% of the total soluble protein in circulating red blood cells.

The entire β-like gene locus has ben sequenced and a large number of structural defects, collectively known as the β-thalassaemias, have been documented in and around the β-globin gene (for review, see Collins & Weissman (1984); Poncz *et al.* (1988)). In a related condition, hereditary persistence of foetal haemoglobin (HPFH), γ-globin gene expression and hence HbF (foetal haemoglobin) production persist into adult life. These diseases are not only clinically important but they also provide natural models for the study of transcriptional regulation during development (see below).

2. THE β GLOBIN LCR

The entire locus is regulated by the locus control region (LCR) which first became apparent from the study of a Dutch thalassaemia (van der Ploeg *et al.* 1979) and later a Hispanic thalassaemia (Driscoll *et al.* 1989). The Dutch thalassaemia contained an intact β globin gene but had a deletion of the upstream part of the locus which removed the LCR and prevented activation of the β gene (Kioussis *et al.* 1983; Wright *et al.* 1984; Taramelli *et al.* 1986). The LCR is characterized by a set of developmentally stable, hypersensitive sites, 5′ HS1, 2, 3 and 4 and is distributed over approximately 15 kb located upstream of the ε globin gene (figure 1; Tuan *et al.* 1985; Grosveld *et al.* 1987; Forrester *et al.* 1987). The deletion of the LCR in these thalassaemias not only showed that it is required for the expression of the β-like globin genes but also that its presence affects chromatin structure over a distance of at least 150 kb (Forrester *et al.* 1990. However, despite the early data on the Dutch thalassaemia (Kioussis *et al.* 1983) and the hypersensitive sites (Tuan *et al.* 1985), the functional significance of the LCR only became clear when it was used to drive the expression of the human globin genes in transgenic mice.

When the regulation of the individual globin genes in the absence of the LCR was studied in transgenic mice, it showed that the γ and β globin genes were expressed in a developmental specific manner, albeit at low levels and dependent on the position of integration in the host genome (Magram *et al.* 1985; Chada *et al.* 1986; Kollias *et al.* 1986; Townes *et al.* 1985). When the β globin gene was studied in the context of the complete LCR region in transgenic mice a completely different pattern was observed, the gene was expressed at levels comparable to that of the endogenous mouse globin genes and it became independent of the position of integration in the mouse genome. In addition, the level of expression was

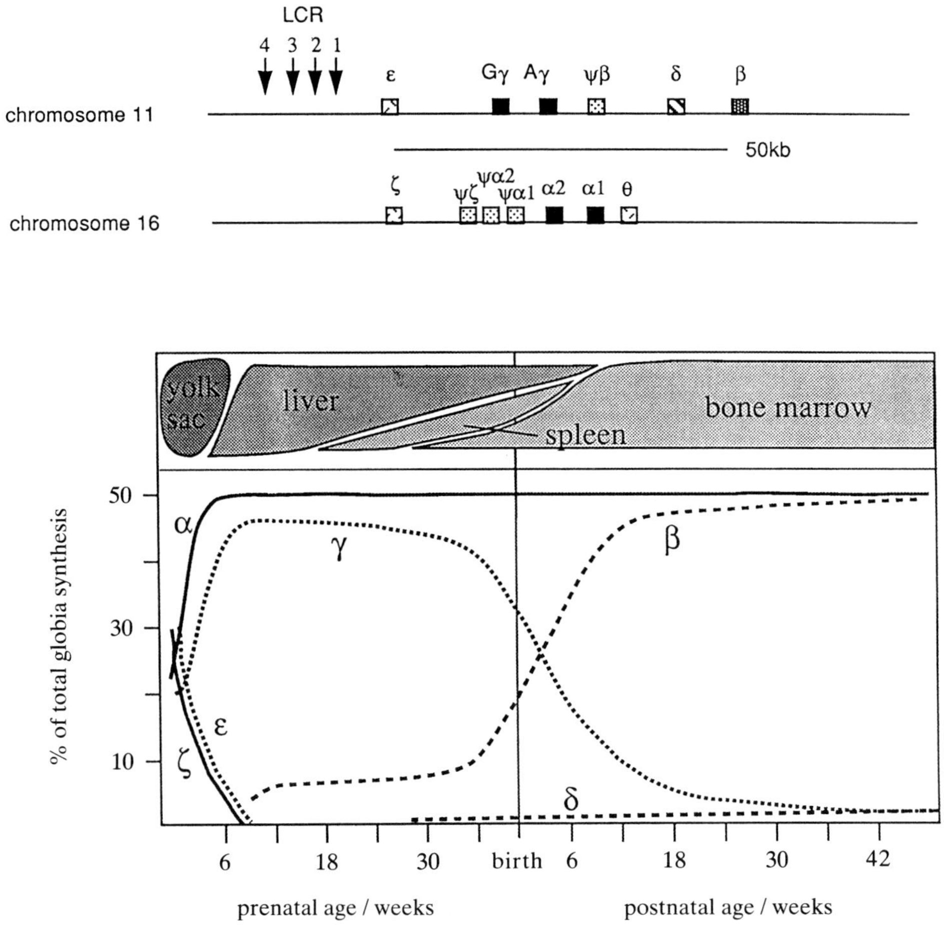

Figure 1. The human α and β loci and the developmental expression of the globin gene expression during human development. Speckled boxes indicate pseudogenes.

dependent on the number of copies of the introduced transgene (Grosveld *et al.* 1987).

3. PROPERTIES OF THE LCR

Position independence and copy number dependence can theoretically be explained by at least two independent mechanisms; either positive activation by the LCR is always achieved and can result in very high levels of expression which obscures small position effects that may still be present in the background, or (and) the region contains elements that insulate it from neighbouring regions, providing a locus border element (LBE) as have been described in Drosophila (Kellum & Schedl 1991). Matrix attachment sites (MAR; Gasser & Laemmli 1986; Jarman & Higgs 1988) or 'A' elements (Stief *et al.* 1989; Bonifer *et al.* 1990) could be LBEs and we initially speculated that these could be part of the LCR in addition to sequences activating sequences (Grosveld *et al.* 1987). However, preliminary experiments indicate that this is not the case and that such a border may be located further upstream (S. Pruzina & F. Grosveld, unpublished data). In addition, small fragments containing any one of the hypersensitive sites can also give rise to copy-number-dependent expression independent of the site of integration albeit at lower levels than the full LCR (Forrester *et al.* 1989; Ryan *et al.* 1989; Talbot *et al.* 1989; Fraser *et al.* 1990). This makes it highly unlikely that insulating sequences form the basis of the effect. It therefore appears that the position independence we observe is (at least in part) due to the fact that the LCR achieves activation of transcription in some dominant fashion perhaps by creating very stable interactions between the LCR and the genes (see below). Consequently positive position effects would only be present as part of the background and only become apparent in situations where the linked gene is suppressed (Dillon & Grosveld 1991; see below for discussion). Interestingly, position effects are not observed when low levels of expression are obtained by the use of inefficient promoters or part of the LCR or mutations in the LCR. This suggests that the stability of the interaction between the LCR and the promoter is not related to the rate of initiation of transcription, except when the promoter is suppressed (Dillon & Grosveld 1991). On the basis of this it may be expected that protein factors exist that are required for the stability of the LCR–promoter complex, in a manner which is separate from the trans-activating function of classical transcription factors. One candidate factor for such a function might be the γ globin promoter factor that increases the competitive ability of the γ gene relative to the β gene in transfection assays (Jane *et al.* 1992; see below for discussion).

4. STRUCTURE OF THE LCR

The main activity of the LCR is associated with HS2, 3 and 4 (Forrester *et al.* 1989; Ryan *et al.* 1989; Tuan *et al.* 1989; Collis *et al.* 1990; Fraser *et al.* 1990; Fraser *et al.* 1992; Lowrey *et al.* 1992), in agreement with the deletion observed in a Hispanic $\gamma\beta$ thalassaemia (Driscoll *et al.* 1989). Each of the sites contains elements capable of activating a linked transgene, independent of the site of integration. A number of erythroid specific and ubiquitously expressed protein binding sites have been mapped to these fragments. Several of specific binding sites are present in all three of the active HS regions (figure 2; Philipsen *et al.* 1990; Talbot *et al.* 1990; Pruzina *et al.* 1991). Very striking are closely spaced binding sites for the erythroid and megakaryocytic specific factor GATA1 (Martin & Orkin 1990; Romeo *et al.* 1990), which has been shown to be essential for erythroid development (Pevny *et al.* 1991). Deletion of GATA-1 binding sites in the promoter prevents erythroid specific induction of the β-globin gene (de Boer *et al.* 1988) and the protein has been shown to have transcriptional activation properties (Martin & Orkin 1990). GATA-1 binds to many erythroid specific genes including all of the globin genes, but the presence of GATA-1 binding sites per se is insufficient to give position independent expression. The immediate 5′ and 3′ flanking regions of the human β globin gene contain at least six GATA-1 binding sites (de Boer *et al.* 1988; Wall *et al.* 1988), but these do not confer integration site independent expression on the β-globin gene (Magram *et al.* 1985; Townes *et al.* 1985; Kollias *et al.* 1986). Our recent experiments with HS3 in transgenic mice indicate that the minimal combination required for position-independent expression is in actual fact not the presence of two closely spaced GATA1 sites, but two GATA1 sites immediately flanking a G rich motif on both sides (Philipsen *et al.* 1992; figure 2). This is in perfect agreement with the binding sites that appear to be occupied *in vivo* (Strauss & Orkin 1992). Interestingly these GATA1 binding sites are completely conserved between the human HS3 and the homologous sequence in the goat (Li *et al.* 1991), which supports our observation on the role of these sites. The G-rich motif in the human and goat HS3 efficiently bind the transcription factor Sp1 (Philipsen *et al.* 1992) *in vitro* suggesting that this factor may play an important role in erythroid specific transcriptional activation. Sp1 would be an attractive candidate because it has been shown to be able to loop DNA (Li *et al.* 1991; Mastrangelo *et al.* 1991; Su *et al.* 1991), a process thought to be central to gene activation in general (for review, see Ptashne (1988)) and to the interactions between the LCR and the globin genes in particular (Hanscombe *et al.* 1991). It should be noted however that neither the human or goat sequence contain the optimal Sp1 binding site (Letovsky & Dynan 1989) and although Sp1 can function through such sites (Li *et al.* 1991), it cannot be excluded that another as yet to be identified factor is important. It is interesting to note that two GATA sites are required for erythroid specific position independent activation which leaves the possibility that a combination of different GATA proteins may be required as part of the activating complex (Yamamoto *et al.* 1990).

A different result is obtained from the detailed functional analysis of HS2. Transient transfection experiments show that classical enhancer activity is associated with this site only (Tuan *et al.* 1989; Ney *et al.* 1990), and not with the others. Dissection of the HS2 showed that a number of proteins are bound to the core fragment (figure 2; Talbot *et al.* 1990). These included the erythroid specific factors NF-E2 (Mignotte *et al.* 1989*a*,*b*) and GATA1 and the ubiquitously expressed factors USF and J-BP (Ellis *et al.* 1993). The NF-E2 binding sites are responsible for the enhancer activity of HS2, but it does not appear to be necessary for position independent globin gene activation (Sorrentino *et al.* 1990; Talbot *et al.* 1990, 1991; Ney *et al.* 1990).

Combinations of deletions of the GATA1 and the ubiquitous factor binding sites of HS2 (USF and J-BP; Talbot & Grosveld 1991; Ellis *et al.* 1993), showed that it was impossible to find any particular binding site combination that was crucial for LCR activity (Ellis *et al.* 1993). However when using these minimal HS2 fragments it became apparent that LCR activity was only obtained when multiple copies of any of the transgene constructs had integrated. This suggests that LCR activity was only restored when these minimal sites could interact with each other, presumably by loop formation. This in turn could explain how the complete LCR might work in vivo, namely by the formation of a loop complex between the LCR elements to form a single larger complex that would

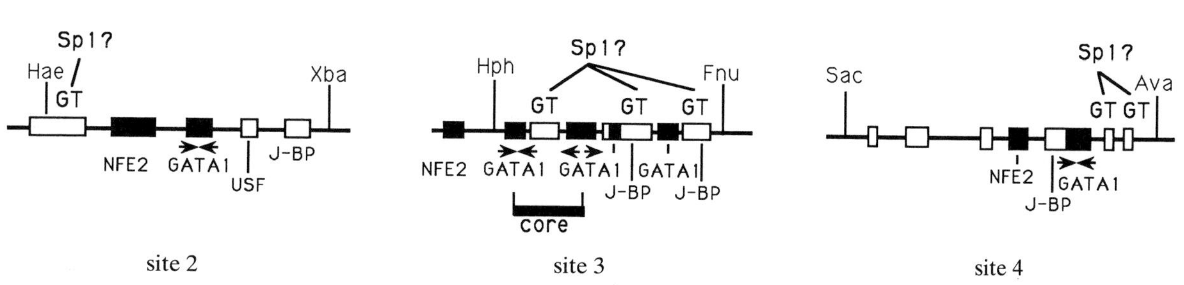

Figure 2. Summary of the protein binding sites HS2, 3 and 4 in the human β globin LCR. Black boxes indicate erythroid specific factor binding, open boxes the binding of ubiquitously expressed factors. The minimal functional core of HS3 is indicated by a thick black line (Philipsen *et al.* 1992).

be capable of interacting with any of the genes. Such a mechanism would also explain why the genes compete with each other for interaction with the LCR, an observation which is difficult to reconcile with a mechanism involving different genes interacting with different elements of the LCR at the same time (see below).

5. DEVELOPMENTAL REGULATION OF THE β-LIKE GLOBIN GENES IN TRANSGENIC MICE

When the developmental regulation of the individual globin genes in the absence of the LCR was studied in transgenic mice, the ε gene was inactive (Shih *et al.* 1990), but the γ and β globin genes were expressed in a developmental specific manner, albeit at low levels and dependent on the position of integration in the host genome (Magram *et al.* 1985; Townes *et al.* 1985; Chada *et al.* 1986; Kollias *et al.* 1986). When the genes were studied in the context of the complete LCR region in transgenic mice, it was found that the ε gene was expressed at the embryonic stage only (P. Fraser, unpublished data) in agreement with the data published for the ε gene linked to part of the LCR (Shih *et al.* 1990; Raich *et al.* 1990). The γ globin gene like its murine structural homologue the βh1 gene, is expressed in the embryonic yolk sac. However in contrast to βh1 the γ gene is expressed in the early foetal liver and is only silenced after day 16 of development. It is not expressed in the adult (Dillon & Grosveld 1991). Although it has not been directly tested in transgenic mice an individual β gene linked to the LCR is expressed prematurely in differentiating embryonic stem cells, although not at maximal levels. Like the murine β gene, the human β gene is expressed at high levels in the foetal liver and adult (Blom *et al.* 1989; Enver *et al.* 1990; Behringer *et al.* 1990; Grosveld *et al.* 1987). This suggests that a large part but not all of the developmental regulation of the globin genes is specified by the regions immediately flanking the genes. However, it is also clear that the LCR is not developmentally neutral and that it influences the expression pattern of the genes (see below).

The simplest model on the basis of these data (see below) was the proposal that stage-specific suppressors are the very important regulators of the individual globin genes (figure 3; Dillon *et al.* 1991). According to this model the LCR can interact with each of the genes at any stage of development but it interacts preferentially with the ε gene in the yolk sac. When the interaction is negated by stage-specific suppressors in the foetal liver the LCR would interact preferentially with the γ globin genes which in turn are suppressed in the adult. At least two candidate binding sites for suppressor factors have been proposed, one in the upstream region of the promoter of the ε globin gene (Cao *et al.* 1989) and one at the distal CAAT box in the promoter of the γ globin genes (Superti Furga *et al.* 1988; Mantovani *et al.* 1989). However, this model does not explain how the later expressing γ and β genes are kept silent at the early stages of human erythroid development. Evidence from natural mutations and transgenic mouse experiments with combinations of genes suggest that competition of the genes for the LCR plays an important role in this process, in particular for the γ and β genes (see below).

6. COMPETITION OF THE GLOBIN GENES IN TRANSGENIC MICE

Competition between the genes became apparent when combinations of genes were used in transgenic mouse experiments. The premature expression of the β gene when it was linked to the LCR and expressed in transgenic mice, could be abrogated by competition for the LCR with another globin gene, e.g. γ or α (Behringer *et al.* 1990; Enver *et al.* 1990; Hanscombe *et*

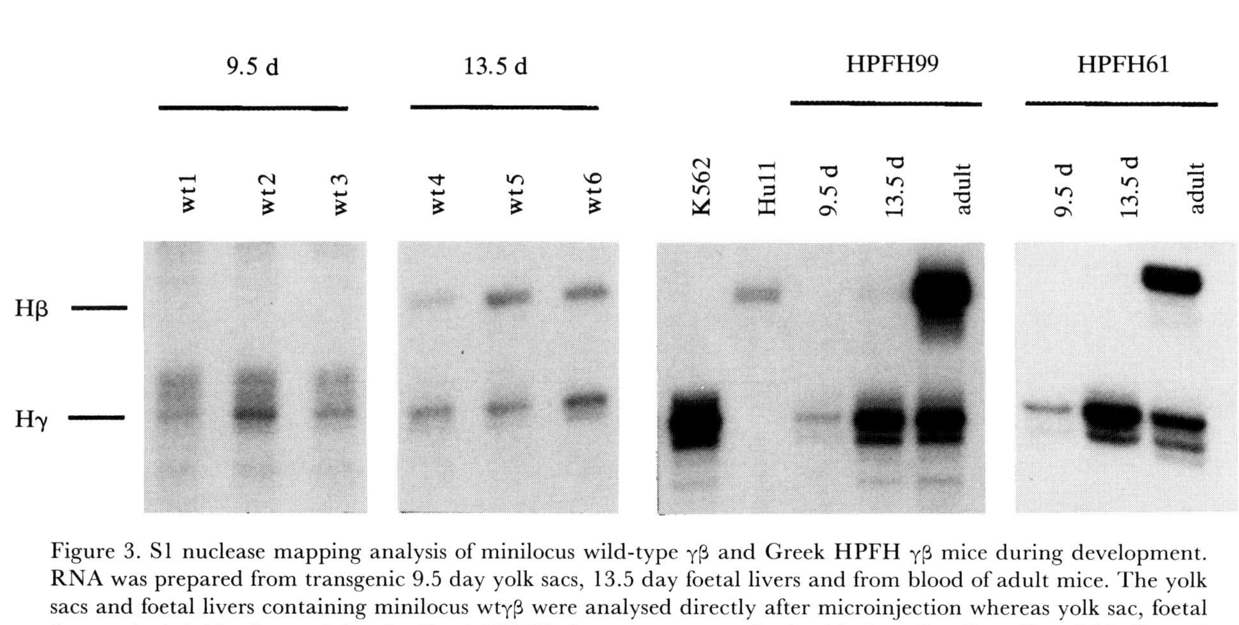

Figure 3. S1 nuclease mapping analysis of minilocus wild-type γβ and Greek HPFH γβ mice during development. RNA was prepared from transgenic 9.5 day yolk sacs, 13.5 day foetal livers and from blood of adult mice. The yolk sacs and foetal livers containing minilocus wtγβ were analysed directly after microinjection whereas yolk sac, foetal liver and adult blood containing the Greek HPFH γβ construct were obtained by breeding lines 61 and 99. Control RNAs were from K562 and Hu11 cells. The signals for human γ- and β-globin RNA are indicated on the left (Hγ, Hβ).

al. 1991). This competition appears to operate in a polar fashion providing an advantage to a gene that is proximal to the LCR and a disadvantage to a gene that is distal to the LCR (Hanscombe *et al.* 1991). As mentioned above these results were subsequently confirmed by the expression of an HPFH (hereditary persistence of foetal haemoglobin) γ and β gene (Berry *et al.* 1992). One group of HPFH mutations is clustered around the distal CAAT box and appear to result in the loss of factor binding sites (Mantovani *et al.* 1989; Fucharoen *et al.* 1990; Gilman *et al.* 1988; Berry *et al.* 1992), suggesting that this region may contain a binding site for a negative regulator. Elevated γ expression in the Greek and Italian HPFHs is associated with a down regulation of the β gene on the same chromosome (Weatherall & Clegg 1981; Giglioni *et al.* 1984). These observations were confirmed by the expression of a γ globin gene which had a single point mutation engineered in the distal CAAT box to mimick the Greek HPFH. Expression of this gene in combination with a β gene in transgenic mice showed that the single point mutation is responsible for the increased expression of the γ gene in the adult and a decrease in expression of the linked β gene. This decrease is particularly strong in the foetal liver due to the increased competitive ability of the γ gene (figure 3; Berry *et al.* 1992).

Why would the expression of the upstream genes prevent expression of those located downstream but not vice versa? Existing competition models are based on the idea that the high level expression of the genes in the β globin locus is potentiated by direct interaction of each gene with the LCR. There is strong evidence that enhancers work through such interactions (Muller *et al.* 1989; Bickel & Pirotta 1990, and references therein). The polar competition which appears to operate in the locus would be explained if the LCR had an intrinsic preference for interaction with a more proximally located gene. A mechanism by which the LCR searches along the chromosome and interacts with the first gene it encounters would be energy requiring and would need very complex machinery. A second possibility would be the formation of chromatin structures which would bring gene and LCR together but these would require specific and developmentally regulated functions for the spacer regions in the locus. A comparison of the phenotypes of different deletions in the locus (reviewed by Poncz *et al.* (1989)), together with the transgenic mouse data, argues against such a role. An alternative and much simpler possibility is that the determining parameter is the relative frequency of contact between genes and LCR (Hanscombe *et al.* 1991). Where two genes each retain a significant capacity to form stable interactions with the LCR, a difference in their frequency of contact with the LCR would dramatically affect competition between them. During the foetal stage the interaction of the γ genes with the LCR would be much more frequent than that of the β gene and, owing to the action of stage specific factors, would also be stronger. The combination of these two parameters acting together would allow the γ genes to compete out β expression. In the adult stage, although the β gene would now have the stronger interaction with the LCR, its much lower frequency of contact would make it difficult for it to compete out a γ gene which retained a significant capability to form such interactions. It is possible that this effect could be achieved by a very strong β interaction but it seems unlikely that this would have evolved specifically to silence γ expression. The more likely alternative mechanism would be one of promoter mediated silencing of the early genes by stage specific factors and the transgenic mouse data indicate that this is the one which has in fact evolved (figure 4). In the case of an HPFH the γ gene would only partially be suppressed and hence retained a significant ability to interact with the LCR, resulting in γ espression and a lower level of β expression. Such a competition mechanism may have evolved to ensure that the switchover from γ to β globin expression is smooth and that the total output of the β gene locus is kept constant and in balance with the output from the α locus.

The model described above would explain the fact that the order of the genes in the β globin locus is largely conserved among mammals. However, it does not predict that the order of expression during development will follow that of the genes in the locus in all species. For example, a gene located proximal to

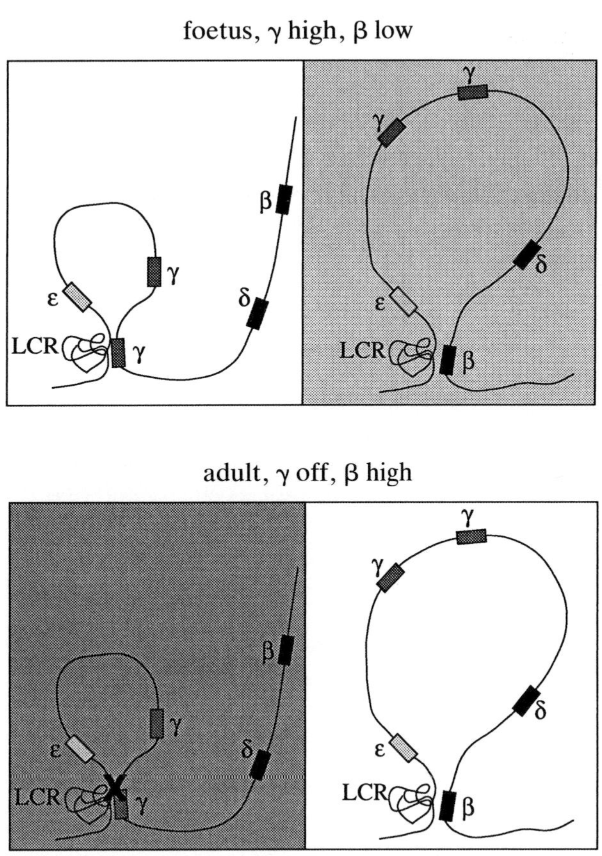

Figure 4. Model of globin gene switching in humans. The genes are indicated by small shaded boxes, high and low levels of expression are indicated by a white or shaded background of each panel. The LCR is symbolized by a large complex and the position of interaction between the LCR and genes are arbitrarily represented. The black cross (adult panel, γ gene) indicates suppression of the γ gene and destabilization of the LCR–γ interaction.

a LCR could be subject to factor mediated repression early in development and then be activated at a later stage. This activation would bring competition into play and would result in a switch from a distal to a proximal gene.

Our experiments with a complete human β globin locus in transgenic mice indicate that such a mechanism may be operating in the early expression period in the mouse. The more distal γ globin genes are expressed and at a higher level than the more proximal ε gene in the yolk sac, which cannot be explained by a simple order–competition model only (Strouboulis *et al.* 1992). Preliminary experiments indicate that the ε and γ genes compete with each other (Fraser *et al.* 1993). This shows that the ε and γ genes behave differently in humans and transgenic mice and suggests that an important change in at least one transcription factor (an ε activator and/or γ suppressor) must have taken place in the process of foetal recruitement of the γ genes in humans (Strouboulis *et al.* 1992). Another possibility to change the patterns of expression of the genes would be to change the position of the LCR. This would also alter the parameters affecting competition. In the chicken β globin locus, for which competition was first proposed (Choi & Engel 1988), there is evidence that part of the LCR lies between the β and ε genes (Reitman *et al.* 1990).

7. DEVELOPMENTAL SPECIFICITY OF THE LCR

The fact that the LCR influences the developmental expression pattern of the genes by competition suggests that the LCR is unable to interact and stimulate transcription of multiple genes at the same time, even though the LCR contains several hypersensitive elements with LCR activity which are distributed over 15 kb of DNA. This could be explained if the individual HS elements would interact with each other (see above for HS2) to form a larger complex, which in turn would interact with the genes. Such a mechanism suggested that the individual HS sites may have a different developmental specificity and that the genes are competing for a specific element in the larger complex at a given developmental stage. We have therefore recently tested whether the individual hypersensitive regions, which make up the complete LCR, interact differently with the genes and whether each region has a different developmental specificity. To answer this question we introduced each hypersensitive region in combination with a set of human γ and β globin genes into transgenic mice and determined the expression pattern of the different human globin genes during development (figure 5; Fraser *et al.* (1993)). The results showed that the individual sites behave differently with the γ or β globin gene, which has a number of implications for the mechanism of LCR function and showed that the LCR is not developmentally neutral but that it influences the expression pattern of the genes. Most striking was the result that HS3 is the only part of the LCR that is capable of directing the expression of the γ globin gene in the foetal liver. This result suggests that this site is the part of the LCR for which the γ and β genes compete during the foetal stage of development. Since all of the other sites are capable of directing β expression during this period this result also indicates that the LCR elements may interact with each other to form a larger single complex (see also above), that interacts with the genes. Interestingly HS4 is the most active site in adult β expression, suggesting that it forms the most stable interaction with the β gene at that stage (Fraser *et al.* 1993). These combination of results indicate that a switch from HS3 to HS4 takes place at the foetal–adult switch and that the γ promoter is capable of forming stable complexes with HS3 but not with HS4 in the LCR complex. As the Greek HPFH gene is expressed effectively in adult mice and able to compete the β gene it suggests that the single point mutation leads to a change in the binding of a factor which positively changes the ability of γ to interact with HS4 in the adult.

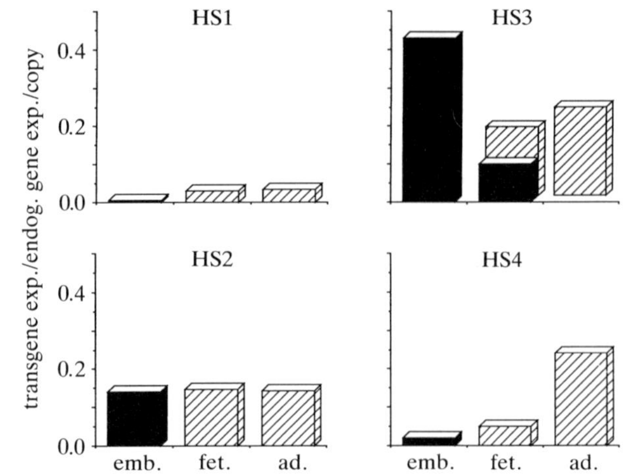

Figure 5. Transgene expression per copy as a percentage of the endogenous mouse genes. Transgene expression levels per copy were calculated by dividing the transgene signal (γ or β) by the endogenous gene signal (βH1 in the embryo; β major in the foetus and adult) and then dividing by 1/2 the transgene copy number (to account for the fact that the mouse has two copies of endogenous genes). Solid columns; γ gene expression. Striped columns; β gene expression.

In summary, the data suggest that the LCR may act through loop formation, possibly achieving activation through a looping between the LCR elements to form a larger single complex which interacts with the genes. The developmental expression pattern of the β globin locus is regulated at different levels. Most important is the action of stage specific regulatory factors on the sequences immediately flanking the genes and perhaps the LCR. In addition, the genes influence each other through a competitive mechanism with the LCR. It is clear that this mechanism involves a number of parameters which include the spatial distribution of the genes relative to the LCR and the stability of the interactions between individual promoters and each of the hypersensitive sites, although the relative contribution of each of these is as yet unknown.

This work was supported by the MRC (U.K.).

REFERENCES

Behringer, R.R., Ryan, T.M., Palmiter, R.D., Brinster, R.L. & Townes, T.M. 1990 Human γ to β globin gene switching in transgenic mice. *Genes Dev.* **4**, 380–389.

Berry, M., Dillon, N. & Grosveld, F. 1992 A single point mutation is the cause of the Greek form of Hereditary Persistance of Foetal Haemoglobin. *Nature, Lond.* **358**, 499–502.

Blom van Assendelft, G., Hanscombe, O., Grosveld, F. & Greaves, D.R. 1989 The β-globin domain control region activates homologous and heterologous promoters in a tissue-specific manner. *Cell* **56**, 969–977.

Bickel, S. & Pirotta, V. 1990 Self association of the *Drosophila* zeste protein is responsible for transvection effects. *EMBO J.* **9**, 2959–2967.

Bonifer, C., Vidal, M., Grosveld, F. & Sippel, A.E. 1990 Tissue specific and position independent expression of the complete gene domain for chicken lysozyme in transgenic mice. *EMBO J.* **9**, 2843–2848.

Cao, S., Gutman, P.D., Dave, H.P.G. & Schechter, A.J. 1989 Identification of a transcriptional silencer in the 5′-flanking region of the human ε-globin gene. *Proc. natn. Acad. Sci. U.S.A.* **86**, 5306–5309.

Chada, K., Magram, J. & Costantini, F. 1986 An embryonic pattern of expression of a human fetal globin gene in transgenic mice. *Nature, Lond.* **319**, 685–689.

Choi, O.-R. & Engel, J.D. 1988 Developmental regulation of human β globin gene switching. *Cell* **55**, 17–26.

Collins, F.S. & Weissman, S.M. 1984 The molecular genetics of human hemoglobin. *Prog. Acid Res. molec. Biol.* **31**, 315–462.

Collis, P., Antoniou, M. & Grosveld, F. 1990 Definition of the minimal requirements within the human beta-globin gene and the dominant control region for high level expression. *EMBO J.* **9**, 233–240.

deBoer, E., Antoniou, M., Mignotte, V., Wall, L. & Grosveld, F. 1988 The human beta-globin promoter; nuclear protein factors and erythroid specific induction of transcription. *EMBO J.* **7**, 4203–4212.

Dillon, N. & Grosveld, F. 1991 Human γ-globin genes silenced independently of other genes in the β-globin locus. *Nature, Lond.* **350**, 252–254.

Driscoll, C., Dobkin, C. & Alter, B. 1989 γδβ-Thalassemia due to a de novo mutation deleting the 5′ β-globin gene activation-region hypersensitive sites. *Proc. natn. Acad. Sci. U.S.A.* **86**, 7470–7474.

Ellis, J., Talbot, D., Pruzina, S. & Grosveld, F. 1993 Synthetic 5′ HS2 construct function as a partially active locu control region. *EMBO J.* (In the press.)

Enver, T., Raich, N., Ebens, A.J., Papayannopoulou, T., Costantini, F. & Stamatoyannopoulos, G. 1990 Developmental regulation of human fetal-to-adult globin gene switching in transgenic mice. *Nature, Lond.* **344**, 309–313.

Forrester, W., Takegawa, S., Papayannopoulou, T., Stamatoyannopoulos, G. & Groudine, M. 1987 Evidence for a locus activator region. *Nucl. Acids Res.* **15**, 10159–10177.

Forrester, W.C., Novak, U., Gelinas, R. & Groudine, M. 1989 Molecular analysis of the human beta-globin locus activation region. *Proc. natn. Acad. Sci. U.S.A.* **86**, 5439–5443.

Forrester, W., Epner, E., Driscoll, C., Enver, T., Brice, M., Papayannopoulou, T. & Groudine, M. 1990 A deletion of the human β globin locus activation region causes a major alteration in chromatin structure and replication across the entire β globin locus. *Genes Dev.* **4**, 1637–1649.

Fraser, P., Hurst, J., Collis, P. & Grosveld, F. 1990 DNaseI hypersensitive sites 1, 2 and 3 of the human beta-globin dominant control region direct position-independent expression. *Nucl. Acids Res.* **18**, 3503–3508.

Fraser, P., Pruzina, S. & Grosveld, F. 1993 Each hypersensitive site of the human beta globin locus control region confers a different developmental pattern of expression to the globin genes. *Genes Dev.* (In the press.)

Fucharoen, S., Shimiza, K. & Fukumaki M. 1990 A novel C-T transition within the distal CCAAT motif of the Gγ globin gene in the Japanese HPFH: implication of factor binding in elevated fetal globin expression. *Nucl. Acids Res.* **18**, 5245.

Giglioni, B., Casini, C., Mantovani, R., Merli, S., Comp, P., Ottolenghi, S., Saglio, G., Camaschella, C. & Mazza, U. 1984 A molecular study of a family with Greek hereditary persistence of fetal hemoglobin and β-thalassaemia. *EMBO J.* **11**, 2641–2645.

Gilman, J., Mishima, N., Wen, X., Stoming, T., Lobel, J. & Huisman, T. 1988 Distal CCAAT box deletion in the Aγ globin gene of two black adolescents with elevated fetal Aγ globin. *Nucl. Acids Res.* **18**, 10635–10642.

Gasser, S.M. & Laemmli, U.K. 1986 Cohabitation of scaffold binding regions with upstream/enhancer elements of three developmentally regulated genes of D. melanogaster. *Cell* **46**, 521–530.

Grosveld, F., Blom van Assendelft, G., Greaves, D. & Kollias, G. 1987 Position-independent high level expression of the human β-globin gene in transgenic mice. *Cell* **51**, 975–985.

Hanscombe, O., Whyatt, D., Fraser, P., Yannoutsos, N., Greaves, D., Dillon, N. & Grosveld, F. 1991 Importance of globin gene order for correct developmental expression. *Genes Dev.* **5**, 1387–1394.

Jane, S., Paul, P., Vanin, E., Gumucio, D. & Nienhuis, A. 1992 Identification of a stage selector element in the human γ-globin gene promoter that fosters preferential interaction with the 5′ HS2 enhancer when in competition with the β-promoter. *EMBO J.* (In the press.)

Jarman, A.P. & Higgs, D.R. 1988 Nuclear scaffold attachment sites in the human globin gene complexes. *EMBO J.* **7**, 3337–3344.

Kellum, R. & Schedl, P. 1991 A position-effect assay for boundaries of higher order chromosomal domains. *Cell* **64**, 941–950.

Kioussis, D., Vanin, E., deLange, T., Flavell, R.A. & Grosveld, F. 1983 β-globin gene inactivation by DNA translocation in γ-thalassaemia. *Nature, Lond.* **306**, 662–666.

Kollias, G., Wrighton, N., Hurst, J. & Grosveld, F. 1986 Regulated expression of human $^{A}\gamma$-, β-, and hybrid γβ-globin genes in transgenic mice: manipulation of the developmental expression patterns. *Cell* **46**, 89–94.

Letovsky, J. & Dynan, W.S. 1989 Measurement of the binding of transcription factor Spl to a single GC box recognition sequence. *Nucl. Acids Res.* **17**, 2639–2653.

Li, Q., Zhou, B., Powers, P., Enver, T. & Stamatoyannopoulos, G. 1991 Beta globin locus activating regions: conservation of organization, structure and function. *Proc. natn. Acad. Sci. USA* **87**, 8207–8211.

Li, R., Knight, J., Jackson, S., Tjian, R. & Botchan, M. 1991 Direct interaction between Spl and the BPV enhancer E2 protein mediates synergistic activation of transcription. *Cell* **65**, 493–505.

Lowrey, H., Bodine, D. & Nienhuis, A. 1992 Mechanism of DNase I hypersensitivity site formation within the human globin locus control region. *Proc. natn. Acad. Sci. U.S.A.* **89**, 1143–1147.

Magram, J., Chada, K. & Costantini, F. 1985 Developmental regulation of a cloned adult β-globin gene in transgenic mice. *Nature, Lond.* **315**, 338–340.

Mantovani, R., Superti-Fuga, G., Gilman, J. & Ottolenghi, S. 1989 The deletion of the distal CCAAT box region of the Aγ globin gene in black HPFH abolishes the binding of the erythroid specific protein NFE3 and of the CCAAT displacement protein. *Nucl. Acids. Res.* **17**, 6681–6691.

Martin, D.I. & Orkin, S.H. 1990 Transcriptional activation and DNA binding by the erythroid factor GF-1/NF-E1/Eryf 1. *Genes Dev.* **4**, 1886–1898.

Mastrangelo, I., Courey, A., Wall, J., Jackson, S. & Hough, V. 1991 DNA looping and Sp1 multimer links: A mechanism for transcriptional synergism and enhancement. *Proc. natn. Acad. Sci. U.S.A.* **88**, 5670–5674.

Mignotte, V., Eleouet, J.F., Raich, N. & Romeo, P.H. 1989 Cis- and trans-acting elements involved in the regulation of the erythroid promoter of the human porphobilinogen deaminase gene. *Proc. natn. Acad. Sci. U.S.A.* **86**, 6548–52.

Mignotte, V., Wall, L., deBoer, E., Grosveld, F. & Romeo, P.H. 1989 Two tissue-specific factors bind the erythroid promoter of the human porphobilinogen deaminase gene. *Nucl. Acids Res.* **17**, 37–54.

Muller, H., Sogo, J. & Schaffner, W. 1989 An enhancer stimulates transcription in trans when attached to the promoter via a protein bridge. *Cell* **58**, 767–777.

Ney, P.A., Sorrentino, B.P., McDonagh, K.T. & Nienhuis, A.W. 1990 Tandem AP-1-binding sites within the human beta-globin dominant control region function as an inducible enhancer in erythroid cells. *Genes Dev.* **4**, 993–1006.

Ney, P.A., Sorrentino, B.P., Lowrey, C.H. & Nienhuis, A.W. 1990 Inducibility of the HS II enhancer depends on binding of an erythroid specific nuclear protein. *Nucl. Acids Res.* **18**, 6011–6017.

Pevny, L., Simon, M.C., Robertson, E., Klein, W.H., Tsai, S.F., D'Agati, V., Orkin, S.H. & Costantini, F. 1991 Erythroid differentiation in chimaeric mice blocked by a targeted mutation in the gene for transcription factor GATA-1. *Nature, Lond.* **349**, 257–260.

Philipsen, S., Talbot, D., Fraser, P. & Grosveld, F. 1990 The beta-globin dominant control region: hypersensitive site 2. *EMBO J.* **9**, 2159–67.

Philipsen, S., Pruzina, S. & Grosveld, F. 1992 The minimal requirements for activity in transgenic mice of hypersensitive site 3 of the β globin Locus Control Region. (Submitted.)

Poncz, M., Henthorn, P., Stoeckert, C. & Surrey, S. 1988 Globin gene expression in hereditary persistence of fetal hemoglobin and γβ thalassaemia. In *Oxford surveys on eukaryotic genes* (ed. N. McLean), vol. 5, pp. 163–203. Oxford University Press.

Pruzina, S., Hanscombe, O., Whyatt, D., Grosveld, F. & Philipsen, S. 1991 Hypersensitive site 4 of the human beta globin locus control region. *Nucl. Acids Res.* **19**, 1413–1419.

Ptashne, M. 1988 How eukaryotic transcriptional activators work. *Nature, Lond.* **335**, 683–689.

Raich, N., Enver, T., Nakamoto, B., Josephson, B., Papayannopoulou, T. & Stamatoyannopoulos, G. 1990 Autonomous developmental control of human embryonic globin switching in transgenic mice. *Science, Wash.* **250**, 1147–1149.

Reitman, M., Lee, E., Westphal, H. & Felsenfeld, G. 1990 Site independent expression of the chicken β^A globin gene in transgenic mice. *Nature, Lond.* **348**, 749–752.

Romeo, P.H., Prandini, M.H., Joulin, V., Mignotte, V., Prenant, M., Vainchenker, W., Marguerie, G. & Uzan, G. 1990 Megakaryocytic and erythrocytic lineages share specific transcription factors. *Nature, Lond.* **344**, 447–449.

Ryan, T.M., Behringer, R.R., Townes, T.M., Palmiter, R.D. & Brinster, R.L. 1989 High-level erythroid expression of human alpha-globin genes in transgenic mice. *Proc. natn. Acad. Sci. U.S.A.* **86**, 37–41.

Shih, D., Wall, R. & Shapiro, S. 1990 Developmentally regulated and erythroid-specific expression of the human embryonic β globin gene in transgenic mice. *Nucl. Acids Res.* **18**, 5465–5472.

Sorrentino, B., Ney, P., Bodine, D. & Nienhius, A.W. 1990 A 46 base pair enhancer sequence within the locus activating region is required for induced expression of the gamma-globin gene during erythroid differentiation. *Nucl. Acids Res.* **18**, 2721–2731.

Stief, A., Winter, D.M., Stratling, W.H. & Sippel, A.E. 1989 A nuclear DNA attachment element mediates elevated and position-independent gene activity. *Nature, Lond.* **341**, 343–345.

Strauss, E. & Orkin, S. 1992 In vivo protein-DNA interactions at hypersensitive site 3 of the human beta globin locus control region. *Proc. natn. Acad. Sci. U.S.A.* **89**, 5809–5813.

Stroubuolis, J., Dillon, N. & Grosveld, F. 1992 Developmental regulation of a complete 70 kb human β-globin locus in transgenic mice. *Genes Dev.* (In the press.)

Su, W., Jackson, S., Tjian, R. & Echols, H. 1991 DNA looping between sites for transcriptional activation: self association of DNA-bound Sp1. *Genes Dev.* **5**, 820–826.

Superti-Furga, G., Berberis, A., Schaffner, G. & Busslinger, M. 1988 The −117 mutation in Greek HPFH affects the binding of three nuclear factors to the CCAAT region of the γ-globin gene. *EMBO J.* **7**, 3089–3107.

Talbot, D., Collis, P., Antoniou, M., Vidal, M., Grosveld, F. & Greaves, D.R. 1989 A dominant control region from the human beta-globin locus conferring integration site-independent gene expression. *Nature, Lond.* **338**, 352–355.

Talbot, D. & Grosveld, F. 1991 The 5′ HS2 of the globin locus control region enhances transcription through the interaction of a multimeric complex binding at two functionally distinct NF-E2 binding sites. *EMBO J.* **10**, 1391–1398.

Talbot, D., Philipsen, S., Fraser, P. & Grosveld, F. 1990 Detailed analysis of the site 3 region of the human beta-globin dominant control region. *EMBO J.* **9**, 2169–2177.

Taramelli, R., Kioussis, D., Vanin, E., Bartram, K., Groffen, J., Hurst, J. & Grosveld, F.G. 1986 Gamma delta beta-thalassaemias 1 and 2 are the result of a 100 kbp deletion in the human beta-globin cluster. *Nucl. Acids Res.* **14**, 7017–7029.

Townes, T.M., Lingrel, J.B., Chen, H.Y., Brinster, R.L. & Palmiter, R.D. 1985 Erythroid-specific expression of human beta-globin genes in transgenic mice. *EMBO J.* **4**, 1715–1723.

Tuan, D., Solomon, W., Li, Q. & London, I. 1985 The "β-like globin" gene domain in human erythroid cells. *Proc. natn. Acad. Sci. U.S.A.* **82**, 6384–6388.

Tuan, D., Solomon, W., London, I. & Lee, D. 1989 An erythroid specific, developmental stage independent enhancer far upstream of the human β-like globin genes. *Proc. natn. Acad. Sci. U.S.A.* **86**, 2554–2558.

Van der Ploeg, L., Konings, A., Oort, M., Roos, D., Bernini, L. & Flavell, R. 1992 γβ thalassaemia: deletion of the γ and δ genes influences β globin expression in man. *Cell* **38**, 265–273.

Wall, L., deBoer, E. & Grosveld, F. 1988 The human beta-globin gene 3′ enhancer contains multiple binding sites for an erythroid-specific protein. *Genes Dev.* **2**, 1089–1100.

Weatherall, D.J. & Clegg, J.B. 1981 *The thalassaemia syndromes*. Boston: Blackwell Scientific Publications.

Wright, S., deBoer, E., Rosenthal, A., Flavell, R.A. & Grosveld, F.G. 1984 DNA sequences required for regulated expression of the β-globin genes in murine erythroleukaemia cells. *Phil. Trans. R. Soc. Lond.* B **307**, 271–282.

Yamamoto, M., Ko, L.J., Leonard, M.W., Beug, H., Orkin, S.H. & Engel, J.D. 1990 Activity and tissue-specific expression of the transcription factor NF-E1 multigene family. *Genes Dev.* **4**, 1650–1662.

7

Gene regulation of flowering

D. Bradley, R. Carpenter, R. Elliott, R. Simon, J. Romero, S. Hantke, S. Doyle, M. Mooney, D. Luo, P. McSteen, L. Copsey, C. Robinson and E. Coen

SUMMARY

A major change in the development of plants occurs upon floral induction. Meristems in certain positions become organized to form flowers. We are studying this process using a combination of genetic, molecular and physiological approaches in *Antirrhinum*. In particular, we are exploiting transposon-induced mutations in genes controlling early switches in floral development. These mutations cause homeotic and heterochronic phenotypes and three categories of genes have been identified. The first includes *floricaula* (*flo*), which is required to switch inflorescence meristems to a floral state. This gene has been isolated and shown to be expressed transiently in bract, sepal, petal and carpel primordia. The second group of genes controls the identity (and sometimes the number) of organs in a whorl. These genes affect overlapping whorls and their mutant phenotypes suggest a combinatorial model for gene action in determining the fate of floral primordia. Some of the regulatory interactions between these genes have been revealed by studying *cis*- or *trans*-acting mutations which have resulted in ectopic gene expression. Genes of the third category determine the identity of organs within one whorl and thus affect the symmetry of the flower. We propose that the interactions of these homeotic genes control the basic patterns of inflorescence and flower development not only in *Antirrhinum*, but also in a diverse range of plant species.

1. INTRODUCTION

Plants develop by the sequential addition of structures to the main body. These structures originate from lateral meristems initiated on the flanks of the main apical meristem. The apical meristem is an undifferentiated tissue and throughout development it integrates signals to determine the identity and positioning of these lateral meristems. Typically, plants initially undergo a vegetative phase of development whereby leaves and shoots originate from these meristems. At a later stage, floral induction may occur whereby these lateral meristems are induced to form completely different structures, flowers rather than leafy shoots.

A number of genes affecting the identity of these different meristems have now been isolated from a variety of plant species (Coen & Meyerowitz 1991). Our own study focuses on three classes of gene in *Antirrhinum* that upon mutation give homeotic and, in some cases, heterochronic phenotypes. These genes can affect the switch in the type of meristems produced, the identity of primordia arising in floral meristems, or the switching of a meristem from indeterminate to determinate growth.

Antirrhinum majus has many genetical and physiological attributes that make it amenable to study: its large flowers are easily emasculated, crossed and scored, the different phases of vegetative and reproductive development are easily seen and can be modified environmentally, and its well-characterized transposable elements allow genes to be isolated and clonal sectors of wild-type or mutant tissue to be studied through development (Carpenter & Coen 1990). This paper will focus on *Antirrhinum*, although reference will be made to the other well-analysed species *Arabidopsis*, allowing the conservation or divergence of molecular mechanisms to be compared between two taxonomically distant species.

2. DESCRIPTION OF WILD-TYPE PLANT DEVELOPMENT AND MUTAGENESIS STRATEGY

The initial phase of *Antirrhinum* growth involves a vegetative apical meristem that produces pairs of leaves opposite to each other at each node in a decussate phyllotaxis (i.e. each pair is at a right-angle to the previous). After a period of vegetative growth the apex changes to an inflorescence meristem, producing smaller leaves (bracts) in a spiral arrangement, generally with a single bract at each node (figure 1*a*). In the axils of these bracts, floral meristems are initiated which produce four whorls of organs separated by very short internodes, so that consecutive whorls appear adjacent to each other.

Antirrhinum flowers are zygomorphic: they can be divided into mirror-image halves by a single plane. The flower consists of four whorls of organs, numbered from the outside of the flower; thus whorl 1 is

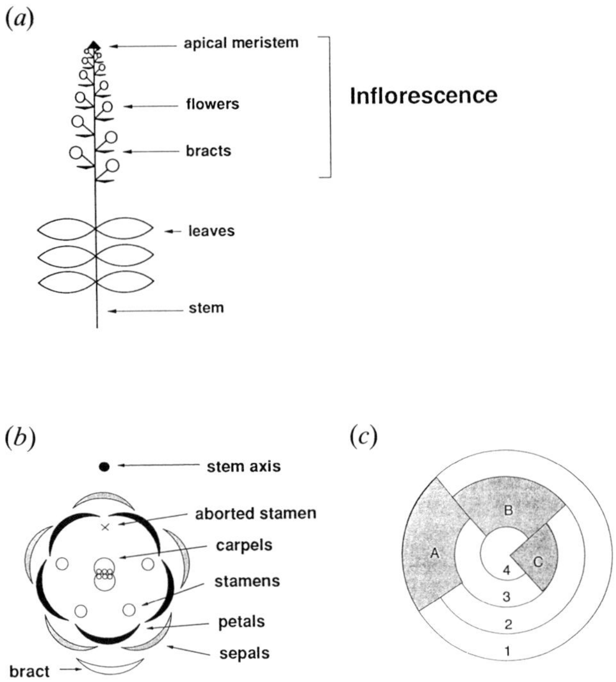

Figure 1. Flowering in *Antirrhinum majus*. (*a*) Schematic illustration of wild-type *Antirrhinum majus* development. (*b*) Floral diagram showing sepal, petal, stamen and carpel organs arranged in four consecutive whorls in the axil of a bract. (*c*) Schematic drawing of the four whorls (1–4) and three regions (A, B and C) of a flower.

outermost and whorl 4 is central (figure 1). The whorl primordia appear sequentially: whorl 1 first, 2 and 3 almost simultaneously and finally whorl 4 (Awasthi *et al.* 1984). Organs within a whorl are upper, if closer to the stem (adaxial), or lower if nearer the bract (abaxial). Whorl 1 consists of five sepals and whorl 2, the corolla, contains five petals which are fused for part of their length, but end in five separate lobes. In whorl 3, five stamen primordia are initiated, though the uppermost fails to develop fully. Whorl 4 is occupied by two united carpels forming a gynoecium with a bilocular ovary (figure 1*b*).

To isolate and study homeotic genes in *Antirrhinum* we have carried out an extensive transposon mutagenesis experiment (Carpenter & Coen 1990). Plants carrying active transposons were grown at 15°C, a temperature inducing transposition in *Antirrhinum* (Carpenter *et al.* 1987), and self-pollinated to give 26 000 progeny, many of which were expected to carry recessive mutations in a heterozygous condition. The homozygous phenotypes were revealed by self-pollination and growing 80 000 plants. Over 15 independent homeotic mutations were obtained from this screen and here we present our conclusions derived from a study of some of the genes involved.

3. GENETIC CONTROL OF THE FIRST STEPS IN FLORAL DEVELOPMENT

The transition from inflorescence to floral meristems represents the first step specific to the floral developmental pathway. Mutants that are unable to carry out this transition might be expected to produce proliferating inflorescence shoots in place of flowers. One mutant obtained with this phenotype is *floricaula* (*flo*) (Carpenter & Coen 1990). The *flo* mutant initiates vegetative growth and the transition to inflorescence meristem in a similar manner to wild-type. However, instead of flowers being produced in the axils of bracts, indeterminate shoots bearing further bracts are produced, each shoot having two opposite bracts at the base followed by a spiral of single bracts (figure 2). Each of these shoots in turn produces further shoots in the axils of their bracts and this sequence is perpetuated. The wild-type *flo* product is therefore necessary for the transition between inflorescence and floral meristems so that in its absence the inflorescence programme is continually reiterated. The *flo* mutant can be considered to be homeotic because it results in one structure (the flower) being replaced by a homologous structure (indeterminate shoot). This mutant can also be viewed as heterochronic (changing the relative timing of events) because it results in an early developmental programme (inflorescence meristem) being continually reiterated.

Other mutants have been described in *Antirrhinum*, such as *squamosa*, which produce inflorescence-like shoots in the axils of bracts, although many of these shoots eventually produce flowers (Stubbe 1966; Huijser *et al.* 1992). Interestingly, although plants with stable *floricaula* alleles never produce flowers, under certain conditions the inflorescence meristem may eventually terminate in carpel-like structures. A similar mutant, *leafy*, has been found in *Arabidopsis* (Haughn & Sommerville 1988; Wiegel *et al.* 1992).

A second class of mutant has almost the opposite effect of those described above because it promotes the conversion of inflorescence to floral meristems in positions where this would not normally occur. Plants with indeterminate inflorescences, such as *Antirrhinum*, do not produce terminal flowers. Thus, the apical inflorescence meristem does not itself undergo the transition to floral. The *centroradialis* mutant in *Antirrhinum* produces terminal flowers, suggesting that the wild-type gene product inhibits this apical conversion (Kuckuck & Schick 1930).

The *flo* gene of *Antirrhinum* has been isolated and extensively characterized (Coen *et al.* 1990). It produces a transcript of about 1.6 kb which has the potential to encode a protein whose sequence shows no extensive homologies with any other proteins in available data-banks, though it contains some motifs found in transcription factors. This suggests that FLO may be a transcriptional activator, although other roles cannot be excluded. RNA analysis reveals that *flo* is induced within two days when plants are shifted to daylength conditions that promote flowering. *In situ* hybridization shows that *flo* is expressed from a very early stage in wild-type inflorescences in a very specific temporal and spatial sequence. The earliest expression seen is in bract primordia and is followed by expression in sepal, petal and carpel primordia, but no expression is seen in stamen primordia. Expression in each organ is transient and is not observed in later stages of development. Taken together, these results

Figure 2. Part of the inflorescence of a *flo* mutant (left) and wild-type (right). All but one of the structures axillary to the bracts have been removed from each plant for clarity.

suggest that *flo* acts not only as a switch between inflorescence and floral meristems, but is also involved in directing specific patterns of gene expression in the early floral meristem.

4. GENETIC CONTROL OF ORGAN IDENTITY

Homeotic mutations were also isolated that act after the inflorescence to floral transtion has taken place. Many of the genes involved affect the identities of organs in two adjacent whorls (Carpenter & Coen 1990; Sommer *et al.* 1990) and for simplicity three overlapping regions (as illustrated in figure 1*c*) can be defined for wild-type: A (whorls 1 and 2); B (whorls 2 and 3); C (whorls 3 and 4). One class of mutant, *ovulata*, affects region A and gives carpels instead of sepals in whorl 1 and stamens in place of petals in whorl 2, giving the overall phenotype carpel, stamen, stamen, carpel.

The second class affects region B and gives sepals instead of petals in whorl 2 and carpels instead of stamens in whorl 3, giving the phenotype sepal, sepal, carpel, carpel (though whorl 4 does not always develop). This class includes the *deficiens* (*def*), *globosa* (*glo*) and possibly some *sepaloidea* (*sep*) mutants. A third class (e.g. *plena*) affects region C and gives petals in place of stamens in whorl 3 and variable structures in whorl 4. Similar classes of whorl identity genes have been found in *Arabidopsis* (Bowman *et al.* 1989; Haughn & Sommerville 1988) suggesting that the mechanisms controlling whorl identity have been highly conserved in evolution. Furthermore, *def* (affecting region B) and *agamous* (an *Arabidopsis* gene affecting region C) share a highly conserved sequence of about 50 amino acids that is also found in yeast and mammalian transcription factors (Schwarz-Sommer *et al.* 1990).

The action of many gene functions in overlapping regions could give each whorl a unique combination of functions. This might provide sufficient information to specify each whorl's identity. However, in order to model these gene interactions it is necessary to introduce some constraints as to how these gene functions are established. The simplest model allows functions active in region B to be established independently of the others, while gene functions of regions A and C are antagonistic (Carpenter & Coen 1990). Consequences of this antagonism have been revealed at the molecular level in *Arabidopsis* and *Antirrhinum*. In wild-type *Arabidopsis* flowers, the gene required for the C function, *agamous*, is expressed only in whorls 3 and 4. However, if the A function is removed, by mutation of *apetala*-2, then *agamous* extends its domain of expression to all whorls (Drews *et al.* 1991). A complementary test for the model is revealed through mutations of *plena*, the *Antirrhinum* homolog of *agamous*, that result in a gain of C function. In this case, certain *cis*-mutations cause ectopic *plena* expression that allows the C function to become established in all whorls, generating a phenotype similar to loss of the A function.

5. REGULATION OF GENES THAT SPECIFY WHORL IDENTITY

Two general types of model can be imagined to explain the expression patterns of genes that determine whorl identity. In a purely spatial model, concentric fields could be set up in the early floral meristem, independently of the sequence of primordium initiation. The gene functions restricted to the defined regions A, B and C could be activated in the appropriate field and hence specify the fate of primordia growing out from the various regions of the meristem. This type of model is formally similar to models proposed for the control of segment identity in *Drosophila* (Ingham 1988). In the sequential type of model, which shares some features of early models of flower development (Heslop-Harrison 1964), the consecutive growth of the primordia would be essential for the establishment of the domains of activity. The various gene functions would be activated in a manner that reflected the sequence of primordium initiation.

Some supporting evidence for the latter sequential model has come from studies on the *flo* gene of *Antirrhinum*. As described above, *flo* is expressed sequentially and appears to be activated in regions where primordia are being initiated. A further feature of *flo* expression is that it occurs transiently in bract, sepal, petal and carpel but not stamen primordia. This pattern suggests that *flo* not only acts to switch inflorescence to floral meristems, but also interacts in a sequential manner with the whorl identity genes that pattern the floral meristem (Coen *et al.* 1990).

6. GENETIC CONTROL OF WHORL NUMBER

Some of the whorl identity mutants also affect whorl number. Extreme mutants in genes affecting region B often give fewer whorls than wild-type. Mutants in genes acting in region C usually have an increase in the number of whorls and a more-or-less indeterminate growth pattern. These findings reveal two functions of the whorl identity genes, namely the control of organ fate and the control of whorl number. The extent to which these separate roles are mediated by the homeotic genes activating particular pathways or target genes is not fully understood.

7. GENETIC CONTROL OF DIFFERENCES WITHIN A SINGLE WHORL

In radially symmetric flowers (actinomorphic), all organs in the same whorl have very similar or identical morphologies, while in zygomorphic flowers (e.g. *Antirrhinum*), one or more whorls have organs which are distinct. The *cycloidea* (*cyc*) mutations in *Antirrhinum* give flowers with a more radially symmetric appearance than wild-type (Stubbe 1966) and four such mutations were isolated in our transposon mutagenesis screen. Extreme *cyc* mutations give flowers with petals all resembling the lowest petal of wild-type. Therefore, *cyc* is a homeotic mutation since organs of one type are replaced by another type. Furthermore, *cyc* appears to affect several whorls, since all stamens develop in *cyc* mutants compared to the abortion of the upper stamen in wild-type.

Does *cyc* affect organs in the same way irrespective of the whorl they occupy? In wild-type, the abortion of the uppermost stamen depends upon *cyc* activity. When stamens are found in whorl 2, in the case of *ovu* mutants, the upper stamens are again aborted, suggesting that *cyc* interacts with primordia in a similar way, irrespective of the whorl they occupy. Thus the fate of a primordium depends on a combination of interactions that determine whorl identity and relative upper or lower position.

8. INFLORESCENCE EVOLUTION

Antirrhinum is among the species with zygomorphic flowers that also have indeterminate inflorescences and so do not normally terminate with a flower. Nevertheless, exceptional plants which produce terminal flowers have been found among these species and they develop terminal flowers with radial symmetry (Peyritsch 1872). This phenotype has been studied in *Antirrhinum* and shown to be due to a recessive allele of the locus *centroradialis*. A few normal flowers are produced laterally in this homeotic mutant, but the apical inflorescence soon produces a radially symmetric terminal flower, with all petals resembling that of the lower petal of wild-type plants, similar to the conversion seen in *cyc*. An explanation for the symmetry of this terminal flower can be found in a polar-coordinate model for zygomorphy (Carpenter & Coen 1990). Floral meristems in axillary positions are in an asymmetrical environment with the main apex above and a bract below them, subjecting them, therefore, to a gradient of *cyc* activity. In contrast, a terminal flower is already in a radially symmetric environment and so may either experience no *cyc* gradient, or no *cyc* activity itself may occur in such an environment.

Mutations such as *centroradialis* also have implications for inflorescence evolution. Some plants may produce a large single flower at the end of their shoots, while in most species flowers are clustered to form inflorescences. In plants with determinate inflorescences, the apical meristem must convert into a floral meristem, with other flowers found in axillary positions or terminating side-branches. An indeterminate inflorescence can produce a quite different array of flowers, as each shoot only produces flowers in axillary positions. The isolation of genes such as *flo*, that regulate the transition from inflorescence to floral meristems, will allow us to address the question of inflorescence evolution at the molecular level. Furthermore, the interactions of *flo* and the organ identity genes should add insight as to how plants overcome the constraints imposed by meristems when trying to pattern and control sequential organ development.

REFERENCES

Awasthi, D.K., Kumar, V. & Murty, Y.S. 1984 Flower development in *Antirrhinum majus* L.(*Scrophulariaceae*) with a comment upon corolla tube formation. *Bot. Mag. Tokyo* **97**, 13–22.

Bowman, J.L., Smyth, D.R. & Meyerowitz, E.M. 1989 Genes directing flower development in *Arabidopsis*. *Pl. Cell* **1**, 37–52.

Carpenter, R. & Coen E.S. 1990 Floral homeotic

mutations produced by transposon-mutagenesis in *Antirrhinum majus*. *Genes Dev.* **4**, 1483–1493.

Carpenter. R., Martin, C.R. & Coen, E.S. 1987 Comparison of genetic behaviour of the transposable element Tam3 at two unlinked pigment loci in *Antirrhinum majus*. *Molec. Gen. Genet.* **207**, 82–89.

Coen, E.S., Romero, J.M., Doyle, S., Elliott, R., Murphy, G. & Carpenter, R. 1990 *floricaula*: a homeotic gene required for flower development in *Antirrhinum majus*. *Cell* **63**, 1311–1322.

Coen, E.S. & Meyerowitz, E.M. 1991 The war of the whorls: genetic interactions controlling flower development. *Nature, Lond.* **353**, 31–37.

Drews, G.N., Bowman, J.L. & Meyerowitz, E.M. 1991 Negative regulation of the *Arabidopsis* homeotic gene *agamous* by the *apetala*-2 product. *Cell* **65**, 991–1002.

Haughn, G.W. & Sommerville, C.R. 1988 Genetic control of morphogenesis in *Arabidopsis*. *Devl Genet.* **9**, 73–89.

Heslop-Harrison, J. 1964 Sex expression in flowering plants. *Brookhaven Symp. Biol.* **16**, 109–125.

Huijser, P., Klein, J., Lonnig, W.-E., Meijer, H., Saedler, H. & Sommer, H. 1992 *Bracteomania*, an inflorescence anomoly, is caused by the loss of function of the MADS-box gene *squamosa* in *Antirrhinum majus*. *EMBO J.* **11**, 1239–1249.

Ingham, P.W. 1988 The molecular genetics of embryonic pattern formation in *Drosophila*. *Nature, Lond.* **335**, 25–34.

Kuckuck, H. & Schick, R. 1930 Die Erbfaktoren bei *Antirrhinum majus* und ihre Bezeichnung. *Z. indukt. Abstamm. u. VererbLehre.* **56**, 51–83.

Peyritsch, J. 1872 Uber Pelorienbildungen. *Sber. Akad. Wiss. Wien.* 66, 1–35.

Schwarz-Sommer, Z., Huijser, P., Nacken, W., Saedler, H. & Sommer, H. 1990 Genetic control of flower development: homeotic genes in *Antirrhinum majus*. *Science, Wash.* **250**, 931–936.

Sommer, H., Beltran, J., Huijser, P., Pape, H., Lonnig, W., Saedler, H. & Schwarz-Sommer, Z. 1990 *Deficiens*, a homeotic gene involved in the control of flower morphogenesis in *Antirrhinum majus*: the protein shows homology to transcription factors. *EMBO J.* **9**, 605–613.

Stubbe, H. 1966 *Genetik und Zytologie von Antirrhinum L. sect Antirrhinum*. VEB. Gustav Fischer Verlag, Jena.

8

New opportunities to dissect and manipulate plant processes

C. R. Somerville

SUMMARY

The use of transgenic plants has become a standard tool of experimental plant biology and is changing many approaches to plant improvement. The technology has greatly expanded the range of methods available to isolate and identify new plants genes, and has permitted great strides in understanding the mechanisms which regulate gene expression. In addition, the ability to use cloned genes to alter the functional expression of the gene in transgenic plants has created entirely novel opportunities to examine the biological role of virtually any cellular constituent.

1. INTRODUCTION

In the decade since the first reports of the production fertile transgenic plants, it has become routinely possible to produce large numbers of transformants of many plant species. For easily transformed species, such as tobacco, the methods for producing transgenic plants are technically simple and inexpensive. Thus, in spite of the fact that several months are usually required to obtain transformed differentiated tissue, production of transgenic plants has become an indispensable component of the methodological repertoire in virtually all aspects of experimental plant biology. For instance, because many methods of gene isolation are based on the use of indirect genetic criteria, the final proof that a gene corresponds to a particular mutation is ultimately the ability of a cloned gene to complement the mutation. Once a gene is cloned, most methods of examining the function of the gene now depend upon the production of transgenic plants.

In this brief overview I have attempted to summarize the various ways in which the methodology is currently exploited, and to speculate on additional possibilities.

2. ISOLATING GENES

(*a*) *Gene-product based cloning*

Insights into many aspects of plant biology can be gained by isolating the genes which control the process of interest. Thus, for instance, in order to understand the mechanisms which regulate the response of plants to light, it has been essential to have access to genes for proteins which regulate (Wagner *et al.* 1991) or are regulated by light (Green *et al.* 1988). However, the utility of the approach is proportional to the ease with which a gene of interest can be isolated. Figure 1 shows schematically most of the methods which are currently employed to isolate plant genes. Until recently, the most widespread method of gene isolation was 'gene product based'. This is typically accomplished by isolating sufficient quantities of a gene product so that an antibody can be raised or so that enough peptide sequence can be obtained to synthesize oligonucleotide probes or polymerase chain reaction (PCR) primers. These materials can then be used to screen appropriate cDNA or genomic libraries, or to amplify target sequences by PCR. The limitation of this approach is that for many genes, the product is not known or cannot be purified in sufficient quantities.

(*b*) *Insertional inactivation*

The currently most popular alternative to product-based cloning is transposon or T-DNA tagging. In those species such as *Antirrihnum majus* or *Zea mays* where several well-characterized transposable elements are known, many important genes have been isolated by tagging (Balcells *et al.* 1991). The limitation of this approach is that it is usually necessary to screen very large numbers of individuals in order to identify a mutation in a particular gene. For instance, in the isolation of the *opaque-2* gene by transposon tagging with the *Spm* element, approximately 500 000 kernels were examined in order to find a mutation. Thus, this approach has been most useful in identifying mutations which result in readily scored phenotypes such as alterations in colour or morphology (Carpenter & Coen 1990).

Since transposable elements are not well characterized in many other important plant species, there has been widespread interest in introducing a well-characterized transposable element into these species. Most of the effort has been focused on the maize Ac/Ds system which appears to retain transposon characteristics when transferred to dicot species such as tobacco

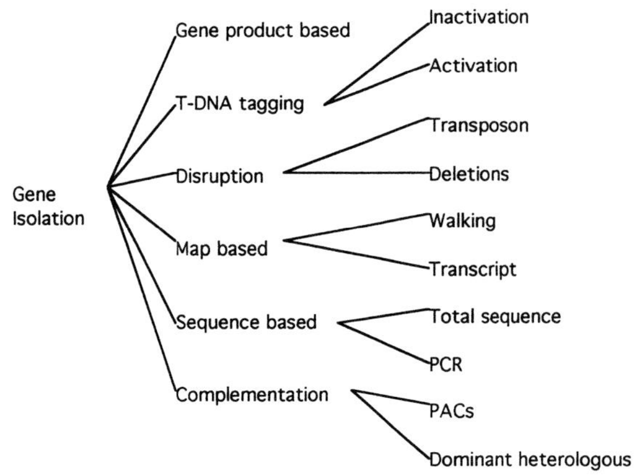

Figure 1. The various methods used in plant gene isolation.

and *Arabidopsis* (Dean *et al.* 1992). Although these heterologous systems have not as yet been sufficiently refined so that they are generally useful, the general approach seems very promising in the long term.

A variant of transposon tagging has been developed to exploit the fact that when transgenic plants are produced, there is inevitably a disruption of the genome by the insertion of ectopic DNA into the chromosome (Feldman 1991; Koncz *et al.* 1990). If the disruption caused a detectable mutation, the disrupted DNA can be readily cloned by using the ectopic DNA as a probe for the flanking sequences. Because it is somewhat tedious to produce transgenic plants on a large enough scale to use T-DNA as an insertional mutagen, the utility of this method has been largely restricted to *Arabidopsis* because it has a small genome which is almost devoid of repetitive DNA. Thus, it appears that a very large proportion of T-DNA insertions generate mutations. Indeed, approximately 19% of all T-DNA insertions produce a visible phenotype (Feldman 1991). In addition to the small genome size, a facile method has been developed for transformation of *Arabidopsis* which has permitted the production of more than 13000 transgenic lines, many of which are available from the *Arabidopsis* Resource Centers at Ohio State University and the University of Nottingham. In this method, seeds are briefly germinated in a culture of *Agrobacterium tumefaciens* then grown to maturity. By some unknown mechanism, a small proportion of the progeny are found to be transformed and can be identified by selection for a kanamycin resistance marker on the T-DNA. Presumably, the plants arising from the seed inoculation become systemically infected and are occasionally transformed by the intracellular population of *A. tumefaciens*. Although the method has led to the isolation of many interesting genes, the approach is inherently limited by the fact that it is opportunistic rather than systematic. That is, if a mutation in a particular gene is not represented in the existing collection, the method is not applicable.

Another promising variant of T-DNA tagging is based on the concept of enhancer trapping. In this method a promoterless gene encoding a reporter gene (e.g. GUS) or selectable marker (e.g. kanamycin resistance) is placed adjacent to a T-DNA border so that if the T-DNA inserts into the genome adjacent to an active promoter, the gene is expressed and the transformed cell survives exposure to the selective agent or expresses the reporter gene (Kertbundit *et al.* 1991; Topping *et al.* 1991). The complementary approach is to place a strong enhancer adjacent to the T-DNA border so that a quiescent gene is activated by the proximity of the enhancer (Walden *et al.* 1991). The large numbers of independent transformation events required for this approach are obtained by protoplast transformation. The key to the successful use of this strategy seems to be in designing screens which will narrow the number of surviving colonies to those in which a particular subset of genes has been targeted. In the first successful application of this method, antibiotic resistant tobacco cells which grow in the absence of exogenous auxin have been identified following transformation with an enhancer. If major regulatory genes control the production of secondary metabolites (see Ludwig *et al.* 1990), this approach might prove very useful for the identification of genes which are required for the production of secondary metabolites such as taxol or other complex plant metabolites of medicinal value. For instance, insertion of an enhancer near the regulatory locus might activate high level expression of the pathway in cultured cells. Many plant-derived drugs are sufficiently complex so that antibodies can be raised against the compounds and these might be used to screen the transformants for the presence of cells which express high levels of the compounds.

(*c*) *Subtractive methods*

The conceptual complement to the isolation of genes on the basis of a transposable element or T-DNA insertion is to exploit the existence of a deletion of a target gene. The utility of this approach was demonstrated by the isolation of a gene which complemented a mutation of the *ga1* locus of *Arabidopsis* (Sun *et al.* 1992). The mutation had previously been suggested to be a deletion on the basis of conventional genetic criteria. A portion of the gene was highly enriched by several successive rounds of hybridization of small fragments of DNA from the wild type with biotinylated DNA of the mutant, followed by removal of DNA hybrids by chromatography on an avidin matrix. The main limitation to the utility of this approach is the difficulty associated with producing and identifying small deletions in higher plants. A variant of the method can, however, be exploited to enrich for genes which are more highly expressed under one condition than another.

(*d*) *Map-based cloning*

The ideal method for gene isolation should be capable of isolating any gene for which a phenotypically detectable recessive variant allele exists. Map-based cloning methods such as chromosome walking are theoretically suitable in that they permit the isolation of any allele which can be genetically mapped. This effectively excludes only quantitative trait loci (QTLs). However, these methods may be considered less than ideal because of the amount of effort required to implement map based cloning in higher plants.

The strategy in map-based cloning is to genetically map an allele of a target gene near previously isolated DNA fragments, typically restriction fragment length polymorphism (RFLP) markers. Then, the region of the genome containing the gene is isolated, and subfragments are used to complement the mutation. *Arabidopsis* is the most suitable plant for the application of map-based cloning methods because, in addition to suitable genetic characteristics, it has a uniquely small nuclear DNA content which is almost devoid of interspersed highly repetitive DNA (Meyerowitz 1989). Several RFLP maps have been constructed (Chang *et al.* 1988; Nam *et al.* 1989), which now contain approximately 312 markers distributed over the five chromosomes. As the total DNA content has been estimated to be 70 000 kb, the average distance from any gene to the nearest RFLP marker is about 225 kb. This is comparable to the average insert size of about 160 kb in the several yeast artificial chromosome (YAC) genomic libraries which are available for *Arabidopsis* (Grill & Somerville 1991; Ward & Jen 1990).

The general utility of these resources was recently demonstrated by the map-based isolation of a gene which complements a mutant (*fad3*) of *Arabidopsis* deficient in omega-3 fatty acids (Arondel *et al.* 1992). In this instance, a YAC covering the region of the *fad3* gene was used to probe a cDNA library prepared from a tissue in which the biological information had suggested the *fad3* transcript would probably be moderately abundant. The YAC encoded only one moderately abundant cDNA. Production of transgenic plants which expressed this cDNA resulted in genetic complementation of the *fad3* mutation, thereby establishing the identity of the gene. In those instances where the target gene is weakly expressed, it is necessary to subclone the YACs and transform the mutant with a series of overlapping subclones to find the complementing fragment. This was essentially the approach which was used to clone the *ABI3* gene from *Arabidopsis* by map-based methods (Giradaut *et al.* 1992). Although it remains somewhat tedious to isolate genes by this approach, it seems likely that with the development of an ordered set of YACs for the *Arabidopsis* genome it will become much easier to implement this approach. At present, more than one third of the genome has been covered by identified YACs of known map position (Hwang *et al.* 1991).

Although *Arabidopsis* has unique advantages for map-based cloning, it may be possible to implement the related method of 'chromosome landing' in other plant species, or in *Arabidopsis*, to accomplish similar ends. The idea is to bypass the construction of a very high density RFLP map and to focus instead on identifying polymorphic markers which are very tightly linked to a target gene. This is possible because of the observation that if random 10 base pair (b.p.) oligonucleotides are used as primers in PCR using genomic DNA from polymorphic individuals, the PCR products frequently exhibit polymorphisms designated RAPDs (randomly amplified polymorphic DNA) (Reiter *et al.* 1992). The central concept underlying the use of RAPDs for chromosome landing is that if one pools DNA from a number of phenotypically identical F2 progeny from a cross between two polymorphic parental lines which carry different alleles for a gene of interest, the pooled progeny will show the polymorphisms of both parents for all regions of the genome except the small region surrounding the target gene (Michelmore *et al.* 1991). In this region, the pooled progeny will necessarily have the same chromosomal composition as the mutant parent. It is possible to identify closely linked DNA probes (i.e. RAPDs) by preparing DNA from each parent and from the pooled F2, then screening these three DNA samples for polymorphisms by the PCR-based method. Polymorphic bands which are present in the pooled F2 and the mutant parent represent fragments of DNA which have a high probability of being closely linked to the mutation. Tight linkage can be verified by scoring additional F2 progeny and tightly linked polymorphic fragments can then be used to screen a YAC library. Because of the relative ease with which a thousand or more different primers can be screened, this method has the inherent potential to generate very tightly linked markers even for species with relatively large genomes.

(*e*) *Large-scale sequencing*

Although not, as yet, widely employed in plant biology, the recent improvements in instruments for large-scale automated DNA sequencing promise to significantly change many current practices. It has now become possible for one laboratory to obtain partial nucleotide sequence (ESTs, expressed sequence tags) for many thousands of cDNA clones in a period of a few months (Adams *et al.* 1991, 1992). Furthermore, because of the large amount of sequence information available in public nucleotide and protein sequence databases, a significant proportion of randomly sequenced cDNA clones can be assigned probable function by comparison to the sequence databases. In the case of human cDNA clones, approximately 18% of the first 5000 ESTs were assigned probable function (Adams et al 1991). Similarly, of the first 1000 ESTs from rice, approximately 20% were assigned probable function: mostly by reference to non-plant genes (Uchimiya *et al.* 1992). In view of the efficiency of this approach as a mechanism for connecting plant biology to the large amount of sequence information available in other orders, it is apparent that very large numbers of ESTs will soon be available for several plants such as rice, maize and *Arabidopsis*.

For the foreseeable future, it will not be possible to assign function to most ESTs by reference to the databank, or the proposed homology will be of marginal statistical significance, or the correspondence between the function of the gene product in one organism and another may be obscured by species specific differences. In these instances, the ESTs will serve essentially as stimuli for hypotheses about the existence of a previously unidentified gene. These hypotheses may be expected to become increasingly important leads into novel aspects of plant biology as more and more genes of readily evident function are characterized. The ability to analyse these ESTs in transgenic plants will be the essential technique for testing these hypotheses.

(*f*) *Cloning by complementation*

Much of the power of yeast genetics is based on the ease with which a gene which complements a mutation can be isolated. Because of the technical difficulties associated with producing large numbers of transformed plants, it is unlikely that the isolation of genes by complementation of plant mutations, per se, will be implemented in higher plants. However, the use of map-based cloning methods is a compromise solution in which the amount of DNA which must be introduced into the mutants is reduced so that fewer transgenic plants need to be produced.

An increasingly important method of cloning plant genes is the use of plant cDNA clones to complement microbial mutants. Many biochemical pathways are similar or identical in plants, bacteria and fungi, and functional complementation of auxotropic mutations is not uncommon (Minet *et al.* 1992). An informative example of the possible opportunities was the recent cloning of two different genes encoding potassium channels by the complementation of the same yeast mutant (Anderson *et al.* 1992; Sentenac *et al.* 1992).

Many of the traits, for which genes are sought, occur in plants in which conventional genetic methods are not likely to be implemented. For example, if one wanted to identify the genes for the synthesis of a particular secondary metabolite in an exotic plant species, it is unlikely that it would be possible to justify the expenditure of time and effort required to establish a genetic system in a new species even if it was amenable to the approach. A possible solution to this problem may be to develop a facile system for the development of PACs (plant artificial chromosomes). This is not possible at present because no plant centromere has yet been isolated. If it were possible to produce libraries of PACs containing foreign DNA in a readily transformable species such as tobacco or *Arabidopsis*, it may be possible to isolate genes by a kind of 'dominant heterologous complementation' in which genes for particular traits are identified by their expression in a distantly related host. This might be a particularly interesting approach to the isolation of indispensable genes which regulate aspects of growth and development. This might eventually also be a useful approach in the isolation of genes for enzymes involved in the synthesis of complex secondary metabolites.

3. GENES AND BIOLOGY

Cloned plant genes are generally used to examine some aspect of genetic mechanisms, to examine the role of the gene product in the life of the organism or to add, alter or remove a trait of practical importance. The methods by which these objectives are realized are summarized in figure 2 and described below.

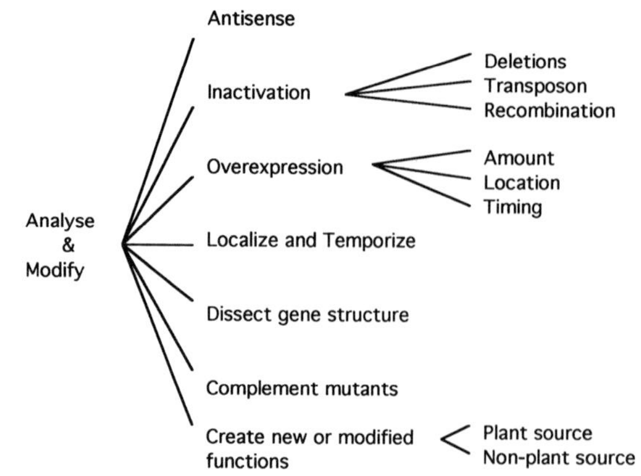

Figure 2. Schematic outline of the ways in which transgenic plants are used to examine the structure and function of plant genes.

(*a*) *Reducing gene expression by antisense or cosuppression*

One of the simplest methods of experimentally reducing the level of accumulation of a gene product is to produce a transgenic plant in which some portion of the gene transcript is expressed in an antisense orientation from an ectopic copy of the gene (Gogarten *et al.* 1992; van der Krol *et al.* 1988). As no firm rules have emerged for the optimal design of an antisense construct, the usual approach is to use the entire transcript, the 5′ half and the 3′ half in three parallel experiments. It is not uncommon to find that all three constructs produce a similar range of effects. As in most experiments with transgenic plants, because of position effects, it is generally necessary to produce a minimum of several dozen independent transgenic plants to as to survey the full range of possible levels of suppression.

Although the mechanism of antisense suppression is not understood, it is possible to imagine mechanisms which account for the effect. By contrast, the observation that expression of ectopic copies of part of a homologous sense transcript may also cause loss of expression of both the ectopic and the native copy of the gene is mysterious. This phenomenon, termed cosuppression, has only been reported in higher plants and has been observed for only a few genes such as chalcone synthase, which is involved in flower pigment formation (Napoli *et al.* 1990; van der Krol *et al.* 1990). In this case the tissue specificity of gene expression can be observed by observing the pattern of pigment formation on the flower. It is apparent from the patterns seen in plants exhibiting cosuppression that the effect may vary depending upon the developmental state of the tissue. The occurrence of colour variegation also indicates that the effect can be unstable as might be expected for an epigenetic phenomenon. The eventual resolution of the mechanism for this interesting effect may eventually suggest novel mechanisms of directed gene control.

(*b*) *Gene inactivation*

Although there have been several reports of homologous replacement of an endogenous gene with a cloned gene, the frequency is too low to be directly useful at present (Halfter *et al.* 1992; Lee *et al.* 1990; Paszkowski *et al.* 1988). However, it has only been several years since similar problems with directed gene replacement in cultured animal cells were overcome by methodological improvements (Capecchi 1989; Mansour *et al.* 1988). The most important innovation was the development of methods for selecting against ectopic integration of exogenous DNA so that only homologous integrations survived selection. It seems likely that the implementation of similar techniques will result in similar improvements in homologous recombination in plant cells. In many plants it is now possible to transform protoplasts at high efficiency and to select rare recombinants. Thus, there seems no intrinsic limitation to the application of this method in easily cultured plant species.

An alternate possibility for obtaining gene disruption (although not gene replacement) may arise from the fact that the Ac/Ds transposon system from maize has been shown to be active in all other plant species tested (see, for example, Dean *et al.* (1992)). The Ac element has the property that it preferentially transposes to a nearby site. This property can, in principle, be exploited by creating a series of transgenic lines of a plant species in which Ac or Ds elements are scattered around the genome. To inactivate a cloned gene, one would first map the cloned gene by RFLP-mapping. In the next stage, a family of Ac/Ds transposition events is produced using an Ac or Ds element which is inserted near the mapped gene. The presence of an individual in the population in which Ac or Ds has jumped into the target gene can be assessed by using the PCR reaction to test for an Ac/Ds insertion in or near the target gene. This can, in principle, be done by preparing total DNA from subpools of all the plants which experienced transpositions, and then using one PCR primer based on the sequence of the Ac/Ds element, and one based on the sequence of the target gene. If the Ac/Ds element lands sufficiently near the target gene, the PCR with produce a new band. The individual transposition event can be identified by successive subdivisions of the population of transposition events.

A similar use of the PCR reaction to isolate genes which have been deleted or disrupted by other mechanisms may also be imagined. For instance, there are approximately 13 000 lines of *Arabidopsis* carrying random T-DNA insertions. If DNA were made from these lines in pools of 50 or 100 plants, it would be possible to screen the entire population for insertion in a target gene by scoring only a few hundred PCR reactions using one or more primers based on the target sequence and a primer based on the T-DNA sequence. Similarly, one might screen pools of mutagenized plants for a specific deletion by using primers which are derived from sequences which are normally too far apart in the wild-type to give rise to a PCR product.

(*c*) *Altered expression of cloned genes*

In some cases the best way to examine the function of a cloned gene may be to increase the amount of expression or to alter the location or timing of expression of the gene. A beautiful example of the utility of the approach was the demonstration that the *Lc* gene of maize was an autonomous transcriptional activator (Ludwig *et al.* 1990). Introduction of the coding sequence of the *Lc* gene under control of a constitutive promoter into various tissues such as roots which are not normally pigmented resulted in expression of the entire pathway of anthocyanin production. This result indicated that the product of the *Lc* gene is a transcriptional regulator which is able to induce transcription of all the genes in the pathway in cells representing many different developmental states.

In some instances a fresh insight into some aspect of the biology of an organism may be gained by introducing genes from other organisms. For instance, in order to alter the composition of the sugars in the vacuole or the apoplast of photosynthetic tissues, Sonnewald *et al.* (1990) used a yeast invertase gene to target abnormal accumulation of invertase to these cellular locations. The invertase hydrolysed sucrose which was normally present in these compartments. This, in turn, caused informative perturbations in the transport of assimilates and in overall photosynthetic activity.

(*d*) *Localizing and dissecting gene expression*

At present, the standard method for identifying where and when a plant gene is expressed is to fuse the promoter sequence of the gene to the coding sequence for a bacterial β-glucuronidase (GUS) and produce transgenic plants (Jefferson *et al.* 1987). The presence of GUS activity can be conveniently detected by histochemical staining of microscopic sections. The widespread and enthusiastic application of this method has led to the accumulation of a vast amount of information about the cell and tissue specificity of gene expression. In many instances, the application of this method has resolved longstanding problems which could not be resolved by other methods. For instance, it had been known for many years that the leaves of dicots contained a cytoplasmic and a chloroplast isozyme of glutamine synthase, but the physiological significance of these isozymes was unknown. Brears *et al.* 1991 resolved this by using GUS fusions of the two promoters to demonstrate that the two isozymes were expressed in distinct cell types with different physiological functions within the leaf.

In the absence of the ability to produce transgenic cells, the study of the relationship between gene structure and function would necessarily be a kind of archeological discipline in which the wonders of extinct creatures are imaginatively reconstructed from bits of fossilized bone. It has been suggested that shooting gold particles coated in DNA into cells at several thousand feet per second and examining the average activity of a reporter gene is only a small step forward in this respect. Fortunately, it is now routinely possible to produce transgenic plants of many species in which the hypotheses derived from such transient expression assays or other criteria are routinely tested (Schell 1987). Although it is not convenient to produce transgenic plants of certain species, the fact that promoters from one species of plant are generally expressed at the correct place and time in distantly related species has permitted the widespread use of a few species as 'universal' recipients of genes from other species.

(*e*) *Introducing new or improved traits*

The potential applications of genetic engineering methods to the improvement of economically important plant species are too vast to enumerate. Many of the goals of traditional plant breeding seem likely to be accelerated by the application of molecular genetics when the relevant genes are available (Gasser & Fraley 1989). For instance, durable disease resistance is consistently ranked as one of the highest priorities of many breeding programs. Substantial progress towards achieving virus resistance has been achieved by the expression of viral coat proteins in transgenic plants (Beachy *et al.* 1990). Similarly, expression of anti-insecticidal proteins in transgenic plants has provided new mechanisms for resistance to insect pests (Williams *et al.* 1992). This latter application is expected to have a very beneficial effect upon the environment by leading to reduced use of insecticides. A creative application of detailed knowledge of tissue specific gene expression has been the use of a pollen-specific promoter to express ribonuclease in developing pollen (Mariani *et al.* 1990). This causes male sterility, a trait which is very useful in the large-scale production of hybrid seed.

Many of the conceivable applications involve altering the quality or quantity of the economic product in ways that cannot be accomplished by conventional plant breeding. Thus, for instance, potatoes which produce very high levels of starch were produced by introducing a bacterial gene for ADPglucose pyrophosphorylase under control of a tuber specific promoter (Stark *et al.* 1992). Unlike the plant enzyme, which is subject to allosteric regulation by intermediates of the starch biosynthetic pathway, the in vivo activity of the bacterial enzyme was not subject to any form of feedback regulation. Thus the bacterial enzyme was not inhibited by the high concentration of biosynthetic intermediates and, as a result, greater metabolic flux of photosynthate toward starch synthesis was possible. Many other modifications of starch, oils or storage proteins can be envisioned by the similar introduction of one of a few genes.

In addition to providing new tools to facilitate conventional breeding, the ability to move genes from one organism to another provides entirely new possibilities to use plants as sources of new biomaterials. An example is the recent production of biodegradable plastics such as poly-β-D-hydroxybutyric acid (PHB) in plants. This compound, which serves as a storage material in many species of bacteria, normally accumulates as small intracellular granules. PHB is made from acetoacetyl-CoA in two enzymatic steps. When the two bacterial genes involved in the conversion of acetoacetyl-CoA to PHB were expressed in transgenic *Arabidopsis* plants, PHB accumulated in small granules within the plant cells and appeared to have the same properties as PHB which normally accumulates in bacteria (Poirier *et al.* 1992). If adequate levels of production of PHB or related polyalkanoates can be produced in transgenic field crops it might be feasible to satisfy a significant proportion of our needs for this class of materials from a renewable source. Furthermore, the availability of an alternate outlet for agricultural production would have a beneficial effect on the farm economy in the developed world and could reduce or eliminate the need for agricultural subsidies.

4. CONCLUDING REMARKS

Experimental plant biology is in the midst of a scientific and technical revolution which has been made possible by the development of methods for isolating genes and analysing their structure and function in transgenic plants. The explosive growth of information which has resulted is fueling a secondary revolution in the ways in which plant improvement is undertaken. More importantly perhaps, the ability to move genes from other orders of organism into plants and to precisely control the expression of foreign genes will expand the ways in which we utilize plants. However, virtually all of the knowledge which has been gained to date concerns temperate plants. To exploit the full potential of this technology, it is essential that work begin on developing the methods for producing transgenic plants of many of the important tropical species upon which much of the worlds population depends for food and fibre. Some of these species, such as oil palm and coconut, pose technical challenges because of long life cycles, but tremendous opportunities because of the very high productivity of these and related species.

REFERENCES

Adams, M.D., Kelley, J.M., Gocayne, J.D., Dubnick, M., Polymeropoulos, M.H., Xiao, H., Merril, C.R., Wu, A., Olde, B., Moreno, R.F., Kerlavage, A.R., McCombie, W.R. & Venter, J.C. 1991 Complementary DNA sequencing: Expressed sequence tags and the human genome project. *Science, Wash.* **252**, 1651–1656.

Adams, M.A., Dubnick, M., Kerlavage, A.R., Moreno, R., Kelley, J.M., Utterback, T.R., Nagle, J.W., Fields, C. & Venter, J.C 1992 Sequence identification of 2,375 human brain genes. *Nature, Lond.* **355**, 632–634.

Anderson, J.A., Huprikar, S.S., Kochian, L.V., Lucas, W.J. & Gaber, R.F. 1992 Functional expression of a probable *Arabidopsis thaliana* potassium channel in *Saccharomyces cerevisiae. Proc. natn. Acad. Sci. U.S.A.* **89**, 3736–3741.

Arondel, V., Lemieux, B., Hwang, I., Gibson, S., Goodman, H.M. & Somerville, C.R. 1992 Map-based cloning of a gene controlling omega-3 fatty acid desaturation in *Arabidopsis. Science, Wash.* **258**, 1353–1355.

Balcells, L., Swinborne, J. & Coupland, G. 1991 Transposons as tools for the isolation of plant genes. *Trends Biotech.* **9**, 31–37.

Beachy, R.N., Loesch-Fries, S., Tumer, N.E. 1990 Coat-protein mediated resistance against infection. *A. Rev. Phytopathol.* **28**, 451–174.

Benfey, P., Ren, N. & Chua, N.H. 1989 The CaMv35S enhancer contains at least two domains which can confer different developmental and tissue-specific expression patterns. *EMBO J.* **8**, 2195–2202.

Brears, T., Walker, E.L. & Coruzzi, G.M. 1991 A promoter sequence involved in cell-specific expression of the pea glutamine synthetase GS3A gene in organs of transgenic tobacco and alfalfa. *Plant J.* **1**, 235–244.

Capecchi, M.R. 1989 Altering the genome by homologous recombination. *Science, Wash.* **244**, 1288–1292.

Carpenter, R. & Coen, E. 1990 Floral homeotic mutations produced by transposon mutagenesis in *Antirrhinum majus. Genes Develop.* **4**, 1483–1493.

Chang, C., Bowman, J.L., DeJohn, A.W., Lander, E.S. & Meyerowitz, E.M. 1988 Restriction fragment length polymorphism map for *Arabidopsis thaliana. Proc. natn. Acad. Sci. U.S.A.* **85**, 6856–6860.

Dean, C., Sjodin, C., Page, T., Jones, J. & Lister,C. 1992 Behaviour of the maize transposable element Ac in *Arabidopsis thaliana. Plant J.* **2**, 69–81.

Feldman, K. 1991 T-DNA insertion mutagenesis in *Arabidopsis*: mutational spectrum. *Plant J.* **1**, 71–82.

Gasser, C. & Fraley, R.T. 1989 Genetically engineering plants for crop improvement. *Science, Wash.* **244**, 1293–1299.

Giraudat, J., Hauge, B., Valon, C., Smalle, J. Parcy, F. & Goodman, H.M. 1992 Isolation of the *Arabidopsis ABI3* gene by positional cloning. *Pl. Cell* **4**, 1251–1261.

Gogarten, J.P., Fichmann, J., Braun, Y., Morgan, L., Styles, P., Taiz, S.L., DeLapp, K. & Taiz, L. 1992 The use of antisense mRNA to inhibit the tonoplast H^+ ATPase in carrot. *Pl. Cell* **4**, 851–864.

Green, P.J., Yong, M.H., Cuozzo, M., Kano-Murakami, Y., Silverstein, P. & Chua, N.H. 1988 Binding site requirements for pea nuclear protein factor GT-1 correlates with sequences required for light-dependent transcriptional activation of *rbcS-3A* gene. *EMBO J.* **7**, 4035–4044.

Halfter, U., Morris, P.C. & Willmitzer, L. 1992 Gene targeting in *Arabidopsis thaliana. Molec. Gen. Genet.* **231**, 186–193.

Hwang, I, Kohchi, T., Hauge, B., Goodman, H., Schmidt, R., Cnops, G., Dean, C., Gibson, S., Iba, K., Lemieux, B., Arondel, V., Danhoff, L. & Somerville, C.R. 1991 Identification and map position of YAC clones comprising one third of the *Arabidopsis* genome. *Plant J.* **1**, 367–374.

Jefferson, R.A., Kavanagh, T.A. & Bevan, M.W. 1987 GUS fusions: β-glucuronidase as a sensitive and versatile gene fusion marker in higher plants. *EMBO J.* **6**, 3901–3907.

Keitbundit, S., de Greve, H., Debroeck, F., van Montagu, M. & Hernalsteens, J.P. 1991 In vivo random β-glucuronidase gene fusions in *Arabidopsis thaliana. Proc. natn. Acad. Sci. U.S.A.* **88**, 5212–5216.

Koncz, C., Mayerhofer, R., Koncz-Kalman, Zs., Nawrath, C., Reiss, B., Redei, G.P. & Schell, J. 1990 Isolation of a gene encoding a novel chloroplast protein by T-DNA tagging in *Arabidopsis thaliana. EMBO J.* **9**, 1337–1346.

Lee, K.Y., Lund, P., Lowe, K. & Dunsmuir, P 1990 Homologous recombination in plant cells after *Agrobacterium* mediated transformation. *Pl. Cell* **2**, 415–425.

Ludwig, S.R., Bowen, B., Beach, L. & Wessler, S.R. 1990 A regulatory gene as a novel visible marker for maize transformation. *Science, Wash.* **247**, 449–450.

Mansour, S.L., Thomas, K.R. & Capecchi, M.R. 1988 Disruption of the proto-oncogene int-2 in mouse embryo-derived stem cells. *Cell* **51**, 503–512.

Mariani, C., Beuckeleer, M., Truettner, J., Leemans, J. & Goldberg, R.B. 1990 Induction of male sterility in plants by a chimeric ribonuclease gene. *Nature, Lond.* **347**, 737–741.

Meyerowitz, E.M. 1989 Arabidopsis, a useful weed. *Cell* **56**, 263–269.

Michelmore, R.W., Paran, I. & Kesseli, R.V. 1991 Identification of markers linked to disease-resistance genes by bulked segregant analysis: A rapid method to detect markers in specific genomic regions by using segregating populations. *Proc. natn. Acad. Sci. U.S.A.* **88**, 9828–9832.

Minet, M., Dufour, M.-E. & Lacroute, F. 1992 Complementation of *Saccharomyces cerevisiae* auxotrophic mutants by cDNAs, *Plant J.* **2**, 417–422.

Nam, H.G., Giraudat, J., den Boer, B., Moonan, F., Loos, W., Hauge, B., & Goodman, H.M. 1989 Restriction fragment length polymorphism linkage map of *Arabidopsis thaliana. Pl. Cell* **1**, 699–705.

Napoli, C., Lemieux, C. & Jorgensen, R. 1990 Introduction of a chimeric chalcone synthase gene into petunia results in reversible co-suppression of homologous genes in trans. *Pl. Cell* **2**, 279–289.

Paszkowski, J., Baur, M., Bogucki, A. & Potrykus, I. 1988 Gene targeting in plants. *EMBO J.* **7**, 4021–4026.

Poirier, Y.P., Dennis, D.E., Klomparens, K. & Somerville, C.R. 1992 Production of polyhydroxybutyrate, a biodegradable thermoplastic, in higher plants. *Science, Wash.* **256**, 520–523

Reiter, R.S., Williams, J.G.K., Feldmann, K.A., Rafalski, J.A., Tingey, S.V. & Scolnik, P.A. 1992 Global and local genome mapping in Arabidopsis thaliana by using recombinant inbred lines and random amplified polymorphic DNAs. *Proc. natn. Acad. Sci. U.S.A.* **89**, 1477–1482.

Schell, J. 1987 Transgenic plants as tools to study the molecular organization of plant genes. *Science, Wash.* **237**, 1176–1183.

Sentenac, H., Bonneaud, N., Minet, M., Lacroute, F., Salmon, J.M., Gaymard, F. & Grignon, C. 1992 Cloning by expression in yeast of a plant potassium transport system. *Science, Wash.* **256**, 663–665.

Sonnewald, U., Brauer, M., von Schaewer, A., Stitt, M. & Willmitzer, L. 1990 Transgenic tobacco plants expressing yeast-derived invertase in either the cytosol, vacuole or apoplast: a powerful tool for studying sucrose metabolism. *Plant J.* **1**, 95–107.

Stark, D.M., Timmerman, K.P., Barry, G.F., Preiss, J. & Kishore, G.M. 1992 Regulation of the amount of starch in plant tissues by ADP glucose pyrophosphorylase. *Science, Wash.* **258**, 287–292.

Sun, T., Goodman, H.M. & Ausubel, F.M. 1992 Cloning the Arabidopsis GA1 locus by genomic subtraction. *Pl. Cell* **4**, 119–128.

Topping, J.F., Wei, W. & Lindsay, K. 1991 Functional tagging of regulatory elements in the plant genome. *Development* **112**, 1009–1019.

Uchimiya, H., Kidou, S., Shimazaki, T., Takamatsu, S., Hashimoto, H., Nishi, R., Aotsuka, S., Matsubayashi, Y., Kidou, N., Umeda, M. & Kato, A. 1992 Random sequencing of cDNA libraries reveals a variety of expressed genes in cultured cells of rice (*Oryza sativa* L.). *Plant J.* **2**(6), 1005–1009.

van der Krol, A.R., Lenting, P.E., Veenstra, J., van der Meer, I.M., Gerats, A.G.M., Mol, J.N.M. & Stuitje, A.R. 1988 An antisense chalcone synthase gene in transgenic plants inhibits flower pigmentation. *Nature, Lond.* **333**, 866–869.

van der Krol, A., Mur, L.A., Beld, M., Mol, J.N.M. & Stuitje, A.R. 1990 Flavonoid genes in petunia: addition of a limited number of gene copies may lead to a suppression of gene expression. *Pl. Cell* **2**, 291–299.

Wagner, D., Tepperman, J.M. & Quail, P.H. 1991 Overexpression of phytochrome B induces a short hypocotyl phenotype in transgenic Arabidopsis. *Pl. Cell* **3**, 1275–1288.

Walden, R., Hayashi, H. & Schell, J. 1991 T-DNA as a gene tag. *Plant J.* **1**, 281–288.

Ward, E.R. & Jen, G.C. 1990 Isolation of single-copy-sequence clones from a yeast artificial chromosome library of randomly sheared *Arabidopsis thaliana* DNA. *Pl. molec. Biol.* **14**, 561–568.

Williams, S., Friedrich, L., Dincher, S., Carozzi, N., Kessman, H., Ward, E. & Ryals, J. 1992 Chemical regulation of *Bacillus thuringiensis* endotoxin expression in transgenic plants. *Biotechnology* **10**, 540–542.

9

Genome manipulation in embryonic stem cells

J. Rossant, C. Bernelot Moens and A. Nagy

SUMMARY

Embryonic stem (ES) cells derived from pluripotent cells of the early mouse embryo provide a powerful tool for genome manipulation in mammals. Dominantly acting effects can be achieved by introducing constructs to misexpress or ectopically express a gene product, express an altered product or express antisense constructs. Use of ES cell chimeras to analyse the effects of such alterations may provide information not readily available from transgenic mice. However, the most important use of ES cells, to date, is in the generation of recessive mutations, either in known genes by targeted mutagenesis or randomly by insertional mutagenesis. Examples of these approaches and possible future strategies are discussed.

1. INTRODUCTION

Successful application of a genetic approach to studying any biological question involves integrating information from the phenotypes of specific mutations with knowledge of the nature of the gene product. In mammals, the mouse has provided a storehouse of useful mutations, accumulated spontaneously or as a result of radiation and other mutagenic strategies. These mutations have included phenotypes indicative of roles for the wild-type genes in development, neurological function, immunobiology as well as models of human genetic diseases. However, few of the genes underlying these existing mutations have yet been cloned. The number of mutations available is still limited, and although progress in this area is accelerating with the advent of large-scale genome characterization, the pace of molecular cloning has meant that there are many more cloned genes for which there is no known mutant phenotype, than there are mutations for which the gene product is known. Recent advances in manipulating the mouse genome have opened up new possibilities for investigating gene function, based on directed mutagenesis or novel insertional mutagenic strategies.

Although introduction of DNA constructs into transgenic mice has been an important advance in genetic manipulation, a major technical breakthrough came with the advent of embryonic stem (ES) cells. In 1981, two groups independently described the derivation of permanent cell lines from the inner cell mass of the blastocyst (Martin 1981; Evans & Kaufman 1981). These cell lines resembled previously described cell lines derived from teratocarcinomas in their ability to proliferate indefinitely in an undifferentiated state in culture, but differentiate into a variety of cell types either in ectopic sites or under altered culture conditions *in vitro*. Most importantly, however, it was shown that ES cells could contribute efficiently to both somatic and germ line tissues after reintroduction into the blastocyst (Bradley *et al.* 1984). Although there had been occasional reports of germ-line colonization from teratocarcinoma cells (Stewart & Mintz 1981), this system had not proved reproducible enough for widespread use. At the same time as the ES cell technology was being developed, other laboratories, most notably those of Oliver Smithies and Mario Capecchi, were showing that targeted mutagenesis via homologous recombination in mammalian cells was an attainable goal (Smithies *et al.* 1985; Thomas *et al.* 1986). In 1987 it was demonstrated that targeted mutations could be introduced into ES cells in culture (Thomas & Capecchi 1987; Doetschman *et al.* 1988). Germ line transmission of such mutations was first reported in 1989 (Thompson *et al.* 1989), and since that time there has been an explosion of interest in applying this technology to the study of the function of many different sorts of gene products. Targeted mutagenesis is not the only new avenue of genetic investigation opened up by the availability of ES cells. Dominant gain-of-function gene constructs can be introduced into ES cells and their effects studied in chimeras. Identification of novel mutations using efficient insertional mutagens is aided by the ability to screen the ES cells in culture for the appropriate insertions. The future of mouse genetics seems sure to involve the continued development of genomic manipulation in ES cells. In this paper, we describe evidence for the pluripotentiality of ES cells, and then review the kinds of mutagenic strategies, both directed and non-directed, that are currently being applied to ES cells. Finally, some possible future genetic approaches using ES cells will be discussed.

2. PROPERTIES OF ES CELLS

ES cells can be maintained in an undifferentiated state in culture indefinitely if maintained on feeder layers of inactivated fibroblasts or in the presence of leukemia inhibitory factor (LIF) (Smith *et al.* 1988; Sanchez *et al.*, 1991). If removed from feeders or LIF, they will differentiate spontaneously into a variety of cell types, including endoderm, muscle and hemopoietic cell types (Doetschman *et al.* 1985; Sanchez *et al.* 1991). They will change their morphology when treated with agents such as retinoic acid, but the full range of differentiation that they can undertake *in vitro* has not yet been well characterized. Most effort has been placed into trying to drive ES cells down the hematopoietic lineage, in the hope of obtaining permanent cell lines capable of repopulating the adult hematopoietic system (Lindenbaum & Grosveld 1990; Wiles & Keller 1991; Chen 1992). Despite this evidence that ES cells *in vitro* have multiple potentialities, these studies cannot prove that the cells are truly pluripotent like the ICM cells from which they derive (Gardner 1968).

The best test of the potential of ES cells is their ability to colonize tissues in chimeras. Despite the widespread use of ES cells, there are actually few studies which provide a detailed analysis of the distribution of ES progeny in chimeras: most studies focus on coat colour chimerism and the ability of cells to colonize the germ line. Existing studies indicate that ES cells can make widespread contributions to adult and embryonic tissues (Bradley *et al.* 1984; Williams *et al.* 1988; Kadokawa *et al.* 1990) and limited contributions to extraembryonic tissues (Beddington & Robertson 1989). A direct comparison of the mosaicism resulting from introduction of ES versus ICM cells into blastocysts has shown that both cell types have equal potential to contribute to all somatic tissues of the resulting mice (Rossant *et al.* 1992). Any deficiencies in ES contributions to certain tissues were shown to be due to effects of the particular genotype contributions used and not to deficiencies in ES potential. The only real difference between ES cells and ICM cells in this assay was the tendency for an increase in mortality with extensive ES contribution to the chimeras.

A clear-cut demonstration of the pluripotentiality of ES cells and the association between high ES controbution and mortality was provided by experiments in which the development of ES- or ICM-tetraploid chimeras was compared (Nagy *et al.* 1990). Tetraploid mouse embryos can develop readily to the blastocyst stage but post-implantation development is poor (Snow 1975; Kaufman & Webb 1990). Embryos rarely proceed beyond early somite stages. However, in combination with diploid cells, tetraploid cells can support development of extraembryonic lineages (Tarkowski *et al.* 1978). When ES cells were aggregated with tetraploid embryos, about 15% of aggregates proceeded to full term and the foetuses recovered were almost all entirely ES-derived. Tetraploid cells made up all of the trophoblast and endoderm layers of the yolk sacs. ICM-tetraploid aggregates developed similarly.

Production of entirely ES-derived fetuses provides a clear demonstration of the full potential of such cells. However, using established ES cell lines such as D3 and AB-1, no entirely ES-derived fetus was able to survive beyond one or two days after birth although entirely ICM-derived offspring were viable. The reasons for this are still unclear, but recent experiments with a newly established cell line, R1, demonstrate that the effect must be due to genetic or epigenetic alterations accumulating with continued passage in culture. Viable entirely ES-derived offspring were produced from about 10% of ES-tetraploid aggregates when early passages (passages 8–15) of this cell line were used (A. Nagy, R. Nagy, W. Abramow-Newerley, J. Roder and J. Rossant, in preparation). However, later passages produced the same non-viable fetuses as observed with other cell lines. At least one sub-clone derived from passage 14 was also capable of generating viable offspring suggesting that some fully pluripotent cells persist within the ES cell population even at late passage.

These chimera experiments demonstrate that it is valid to study the behaviour of genetically manipulated ES cells in chimeras or tetraploid aggregates as well as studying the effects of genetic alterations after germ line transmission.

3. INTRODUCTION OF GENETIC ALTERATIONS IN ES CELLS

(*a*) *Directed strategies*

Once a gene has been molecularly characterized and both cDNA and genomic clones are available it is possible to devise a number of strategies to address its function in the context of the whole organism. Approaches that do not require disruption of the endogenous gene include overexpression or ectopic expression of an activated form of the protein, and introduction of antisense or dominant negative constructs designed to interfere with endogenous gene function. All of these approaches can be achieved via standard transgenesis, although the ES route may have some advantages in situations where the predicted outcome is embryonic death or gross abnormality. Accumulating such information via analysing midgestation transgenic mice can be slow. However, ES cell lines carrying the dominant-acting construct of interest can be used to make large numbers of diploid or tetraploid chimeras which can then be studied for phenotypic effects. This approach has been used to study the effect of overexpression of v-*src* on embryonic development (Boulter *et al.* 1991) but has not yet been fully exploited.

Most attention has been directed towards the use of ES cells to introduce recessive loss-of-function mutations into genes of interest by homologous recombination. The common strategies for targeted mutagenesis (Capecchi 1989) all involve vectors which disrupt the endogenous allele by introduction of a selectable marker, such as the neomycin resistance

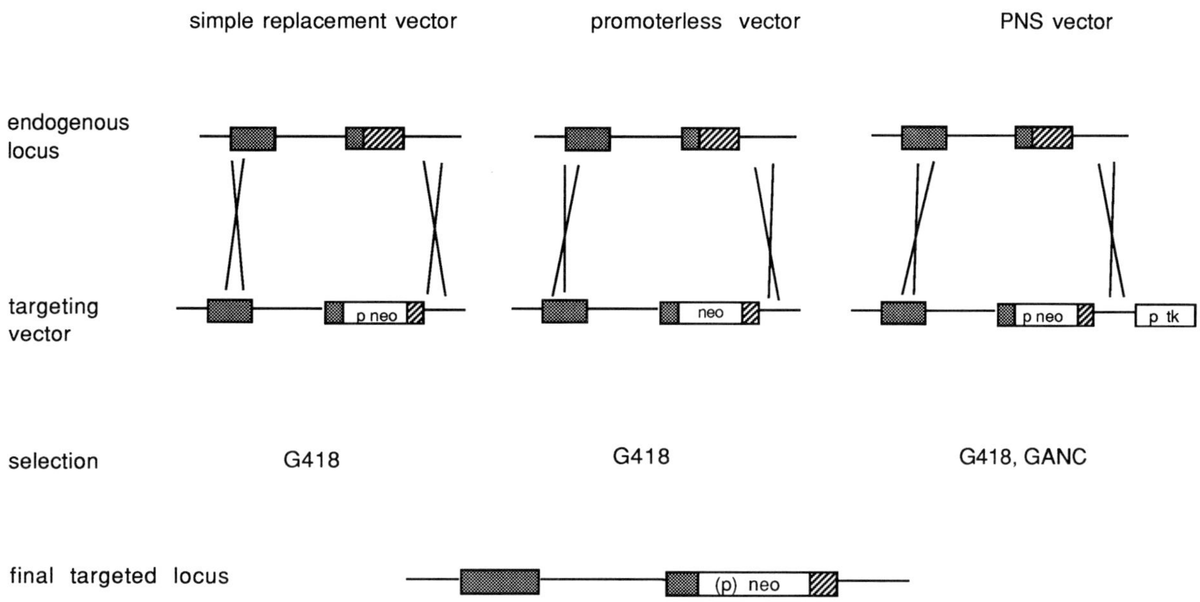

Figure 1. General strategies for targeted mutagenesis using replacement-type vectors. Abbreviations: PNS = positive–negative selection; GANC = gancyclovir; tk = herpes simplex thymidine kinase; neo = neomycin (G418) resistance gene; P = promoter.

gene, into the coding sequence. Identification of a targeted versus non-targeted integration of the targeting vector involves PCR and Southern analysis; enrichment for targeted events can be achieved by positive–negative selection (PNS) (Mansour *et al.* 1988) or promoterless constructs (Charron *et al.* 1990; Te Riele *et al.* 1990) (figure 1). Successful disruption of many genes has now been reported and the phenotypes of such mutations described. It is impossible to review all the results but some general comments can be made.

First, reported targeting efficiencies vary widely, especially when expressed as the number of targeted colonies per *neo*^{r} colonies. Reported efficiencies vary from as high as 1/30 to < 1/5000 for standard PNS vectors. Using promoterless constructs for genes expressed in ES cells, efficiencies as high as 85% have been reported (Te Riele *et al.* 1990). All the reasons for this variation are not clear but some factors have been identified. A relationship between length of homology in the targeting vector and targeting efficiency has been reported (Thomas & Capecchi 1987). There was also some early indication that expressed genes might be easier to target than non-expressed genes. Successful mutagenesis of several genes apparently not expressed in ES cells has tended to argue against the importance of transcriptional activity (Johnson *et al.* 1989). The use of targeting vectors derived from genomic DNA of the same mouse strain as the ES cells used has had a dramatic effect on targeting efficiencies at particular loci. For example, a 20-fold improvement in targeting efficiency was obtained at the *Rb* locus when a vector derived from 129 rather than BALB/c DNA was used (Te Riele *et al.* 1992). This effect is thought to be due to small sequence differences in non-transcribed DNA between mouse strains. The full impact of the effect of using isogenic DNA in targeting vectors has not yet been felt, but the consistently high efficiencies reported with such isogenic vectors suggests that this may be one of the most important factors required for efficient homologous recombination.

Second, phenotypes of targeted mutations have not always been as predicted from knowledge of the nature of the gene product and its pattern of expression. The production of mice lacking cytotoxic T-cells after disruption of the CD8 gene (Fung-Leung *et al.* 1991) and mice lacking mature lymhocytes after discruption of the RAG-1 and RAG-2 V(D)J recombinase genes (Mombaerts *et al.* 1992; Shinkai *et al.* 1992) provide examples where prediction and actual outcome were very close. In other cases, phenotypes appear to reflect a subset of the expression domains of the gene. For example, loss of *Wnt*-1 causes deletion of part of the brain but has no effect on the developing spinal cord where the gene is also expressed (McMahon & Bradley 1990; Thomas & Capecchi 1990). In extreme cases, gene disruption may result in no obvious phenotype, as has been reported for the p53 (Donehower *et al.* 1992) and *PrP* genes (Lang *et al.* 1992) (although p53$^-$ mice are more susceptible to tumours). A full explanation for this rather unpredictable relationship between expression and phenotype cannot yet be provided but is thought to derive from complete or partial redundancies in the gene function either at the single gene level or the genetic pathway level. Sorting out these genetic interactions will require more complete knowledge of genetic pathways and analysis of the effects of many possible combinations of mutations.

All the germ-line mutations described to date have been designed to generate a null mutation in the gene of interest. It is clearly important also to be able to generate more subtle mutations in order to fully dissect gene function. Site-specific mutations within the coding sequence may help elucidate the functions of specific protein domains and mutations in the regulatory sequences may be used to alter gene function in subsets of the gene's normal expression domains. Ideally such mutations should introduce the specific sequence changes without the insertion of a selectable marker gene which could, in itself, affect gene expression. A technique that allows such mutations has recently been described using vectors called 'hit-and-run' (Hasty *et al.* 1991) or 'in–out' (Valancius & Smithies 1991). In this approach an insertion vector containing a positively and negatively selectable marker and the small mutation of interest is introduced into cells. Selection against the marker gene favors the retention of cells in which the vector has recombined out of the cells again, leaving the subtle mutation behind in a proportion of cases. This approach has not yet been extensively used and so the efficiency at different loci is still not clear. It may be possible to devise strategies that favour the 'out' event more strongly by incorporating specific recognition sequences for site-specific recombinases, such as the FLP recombinase from yeast (Golic & Lindquist 1989) and *Cre* (Dale & Ow 1991), into the vectors and causing efficient excision of the marker genes by transient expression of the appropriate recombinase. The use of such recombinases in mammalian cells is just in its infancy (O'Gorman *et al.* 1991) and seems likely to have multiple uses for genome manipulation.

The importance of having multiple alleles at any locus for full understanding of gene function is exemplified by many studies in *Drosophila*, and by a few examples in mammals, such as the *W* locus, where alleles with varying severity contain different site-specific mutations (Reith *et al.* 1990). We have recently shown that a leaky mutation targeted to the N-*myc* locus produces a different phenotype to the null mutation when homozygous (Bernelot-Moens *et al.* 1992). N-*myc* is a nuclear protooncogene in the b-HLH *myc* gene family, and is expressed at high levels in a variety of cell lineages during mouse embryogenesis (Zimmerman *et al.* 1986; Mugrauer *et al.* 1988). There is still very little information on the function of *myc* genes in cells and so there has been considerable interest in mutating them to see if this can shed light on their possible roles in tissue-specific transcriptional regulation. We and others (Charron *et al.* 1990; Stanton *et al.* 1990; Sawai *et al.* 1991) have generated null mutations in the N-*myc* allele *in vitro* using a replacement-type vector that disrupts the coding sequence at the start of translation (figure 2). Mice homozygous for such mutations die in mid-gestation (*ca.* day 9–10) at a stage when elucidation of lineage-specific effects is difficult (B. Stanton & L. Parada, personal communication; J. Charron & F. Alt, personal communication).

We generated another targeted cell line in which the expected gene replacement did not occur but instead the *neo*-containing vector inserted into the first intron of N-*myc* (figure 2). This allows the possible generation of some normal N-*myc* transcripts if mRNA splicing bypasses the *neo*-containing exon. Mice homozygous for this potentially leaky mutation die at birth because of an inability to oxygenate their lungs. This is due to a failure of proliferation of the epithelium during branching morphogenesis of the lungs, indicat-

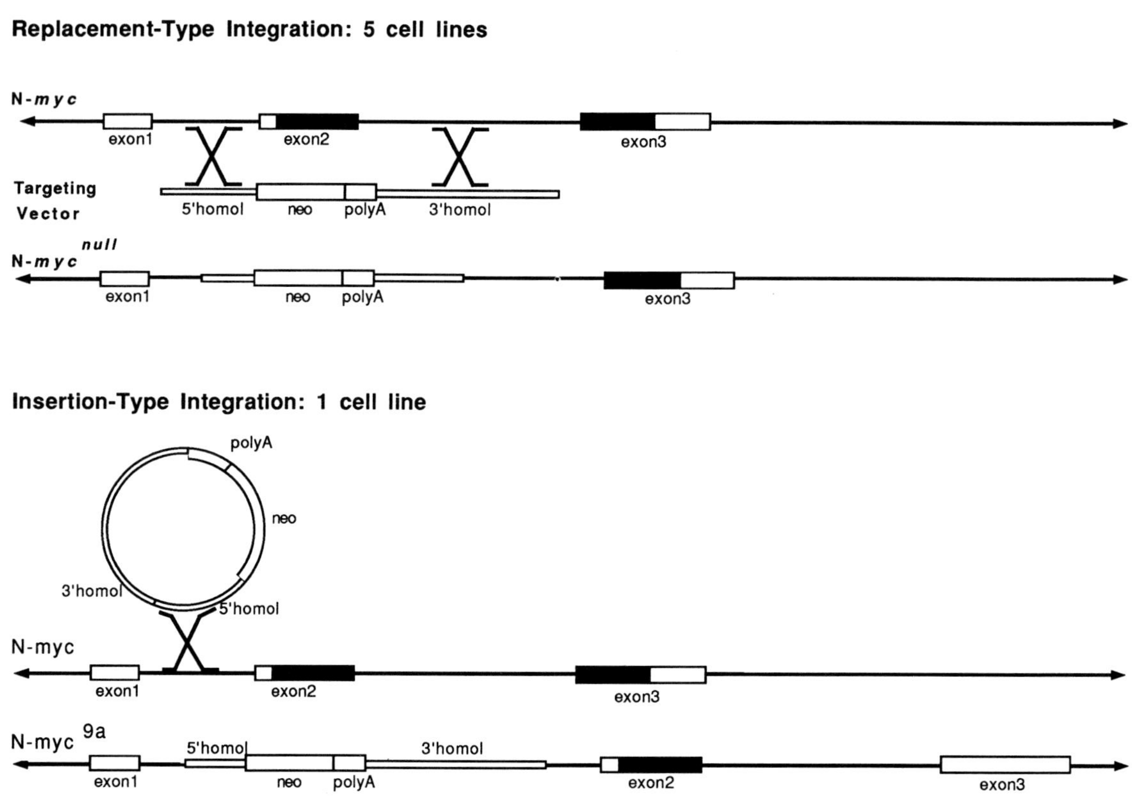

Figure 2. Outcome of gene targeting at N-*myc* locus using a promoterless *neo* vector. In five lines, the targeting vector integrated by the expected double cross-over, replacing exon 2 with the *neo* gene. In one line, the replacement vector recircularized and inserted by a single cross-over in intron 1, resulting in the N-*myc*^{9a} allele shown.

Table 1. *Phenotypes of N*-myc *mutations*

	time of death	phenotype
N-$myc^{null/null}$	9–10 dpc	failure of general cell proliferation
N-$myc^{9a/9a}$	term	failure of epithelial proliferation in lung
N-$myc^{9a/null}$	11–13 dpc	lung defect more pronounced: hypoplasia of myocardium

ing a possible role for N-*myc* in the downstream response to the inductive signals from the lung mesenchyme. Analysis of N-*myc* levels showed that there were some normal N-*myc* transcripts in all tissues analysed, varying from 15–50% of wildtype levels.

The effects of N-*myc* deficiency on lung development could not have been determined from the null mutant mice, since they die before lung morphogenesis has occurred. With this in mind, we reasoned that making double heterozygotes with the null and the leaky mutation might generate an intermediate phenotype and indicate other lineages where N-*myc* is required. Preliminary data from such an experiment indicate that such double heterozygotes die around day 12–13 of development (C. Moens, B. Stanton, L. Parada and J. Rossant, unpublished data). The effects of a further reduction of N-*myc* expression are manifest as an even more severe effect on lung branching but also as a defect in heart development. Specifically the ventricular myocardium is extremely thin and the fetuses appear to die because of inefficient blood circulation. This result indicates an unexpected role for N-*myc* in development of cardiac muscle (table 1).

These results indicate that hypomorphic alleles can be very useful in elucidating gene function at different stages of development. However, insertions of targeting vectors into introns is not a guaranteed way of making leaky or hypomorphic mutations. Use of more directed approaches to generating an allelic series of mutations at specific loci would seem to be an important task for the coming years.

(*b*) *Non-directed strategies*

Targeted mutagenesis can provide insights into the roles of already cloned genes but it is unlikely that all the genes involved in any process are currently available as cloned products. It is important therefore, to continue to search for new genes and mutations affecting any particular process of interest. In species such as *Drosophila* large-scale mutagenic screens have been successfully carried out using efficient chemical mutagens, and sets of mutations affecting different stages of development have been characterized (Jurgens *et al.* 1984; Wieschaus *et al.* 1984). In mouse, the size of the animals and the limited number of offspring make such strategies extemely slow and expensive. Attention has turned towards the use of insertional mutagenesis which, although less efficient than chemical mutagenesis, has the advantage that the inserted DNA sequences can act as a tag for cloning the mutated endogenous gene.

There are a variety of approaches for inducing insertional mutagenesis in mice (Gridley *et al.* 1987). Microinjection of DNA into zygotes (Jaenisch 1988) or retroviral infection of embryos (Jaenisch 1976) has proved to be an effective means of generating novel mutations, and, at least for retroviral insertions, cloning the host gene has been relatively easy (Gridley *et al.* 1991; Jaenisch *et al.* 1983; Weiher *et al.* 1990). The insertion of DNA by microinjection tends to cause large rearrangements of the host DNA which can complicate identification of the mutated gene (Covarrubias *et al.* 1987; Wilkie & Palmiter 1987), but successful cloning of the genes mutated by such transgenic insertions has been reported (Maas *et al.* 1990; Lee *et al.* 1992).

Insertional mutagenesis in ES cells offers several advantages over standard approaches. Using standard insertional mutagens, such as retroviruses (figure 3), it is possible to introduce multiple insertions into single ES cells and populate the germline of chimeras with multiple ES cells (Robertson *et al.* 1986). This results in an increased number of mutagenic events per gamete tested. This approach has been pursued in the laboratory of Dr Martin Evans (Cambridge) and has led to the identification of a number of new mutations (see Conlon *et al.* 1991).

More recently different types of vectors have been used which are designed to allow preselection for integration in or near expressed genes, thus increasing the efficiency of mutagenesis. Three general types of vector have been described, all of which use a reporter gene to allow visualization of gene expression (figure 3). Enhancer trap vectors, containing the *E. coli* β-galactosidase (*lacZ*) gene under the control of a weak promoter, have been introduced into ES cells by electroporation, and expression of *lacZ* has been studied in chimeras (Gossler *et al.* 1989). A recent study of 59 enhancer trap lines has shown that this is a very efficient means of observing novel patterns of gene expression: 1/3 of all lines containing the vector showed *lacZ* expression patterns in chimeras during early development (A. Gossler, personal communication). If these ES cell lines are introduced into the germline they can provide a rich source of novel *in situ* markers for many studies. Presumably because the vector is introduced by electroporation rather than microinjection, the integrations are clean and it has proved possible to identify transcripts closely associated with 3/3 integrations studied so far (A. Gossler, personal communication). However, because integrations are not necessarily within host transcribed sequences, expression of the reporter may not reflect the full expression pattern of the host gene and the integration need not be mutagenic.

Gene- or promoter-trap vectors have more stringent requirements for activation of *lacZ* expression, namely that the integration be within the transcribed sequences of a gene. Promoter-trap vectors contain *lacZ* with an initiation codon but no promoter (Reddy *et al.* 1991). Expression requires insertion into an exon of a gene transcribed in ES cells, either in frame in a translated exon, or in any 5′ untranslated exons. Activation is thus a relatively rare event. A higher

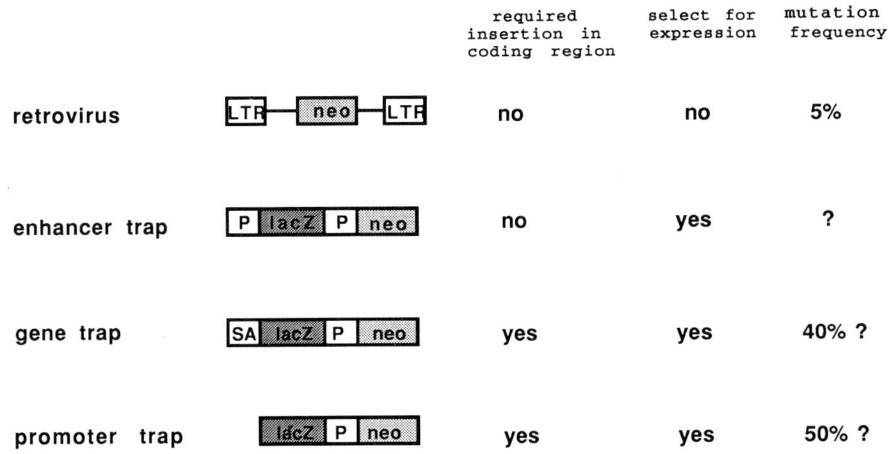

Figure 3. Insertional mutagenesis in ES cells. Abbreviations: LTR = long terminal repeat; P = promoter; SA = splice acceptor.

frequency of activation can be achieved using gene-trap vectors which contain the reporter gene with a splice acceptor sequence upstream (Gossler *et al.* 1989; Brenner *et al.* 1989; Friedrich & Soriano, 1991; Kerr *et al.* 1989). Integrations within introns can allow activation of *lacZ* if there is correct inframe splicing from host exon to the *lacZ* sequence. Both kinds of vectors are likely to cause disruption of host gene transcription and thus be highly mutagenic. This has been confirmed by three recent studies which together suggest that at least 1/3 of all integrations show a readily visible mutant phenotype (von Melchner *et al.* 1992; Friedrich & Soriano 1991; Skarnes *et al.* 1992). Another advantage of these vectors is that expression of the reporter is likely to reflect that of the host gene since *lacZ* is being produced as a fusion transcript with the endogenous gene. Also, cloning the associated exons can be achieved by RACE (Skarnes *et al.* 1992), rather than requiring the cloning of genomic flanking sequences.

All these factors have combined to make the use of these vectors, either introduced by electroporation or in retroviruses, the most efficient means of identifying and mutating novel genes expressed during early

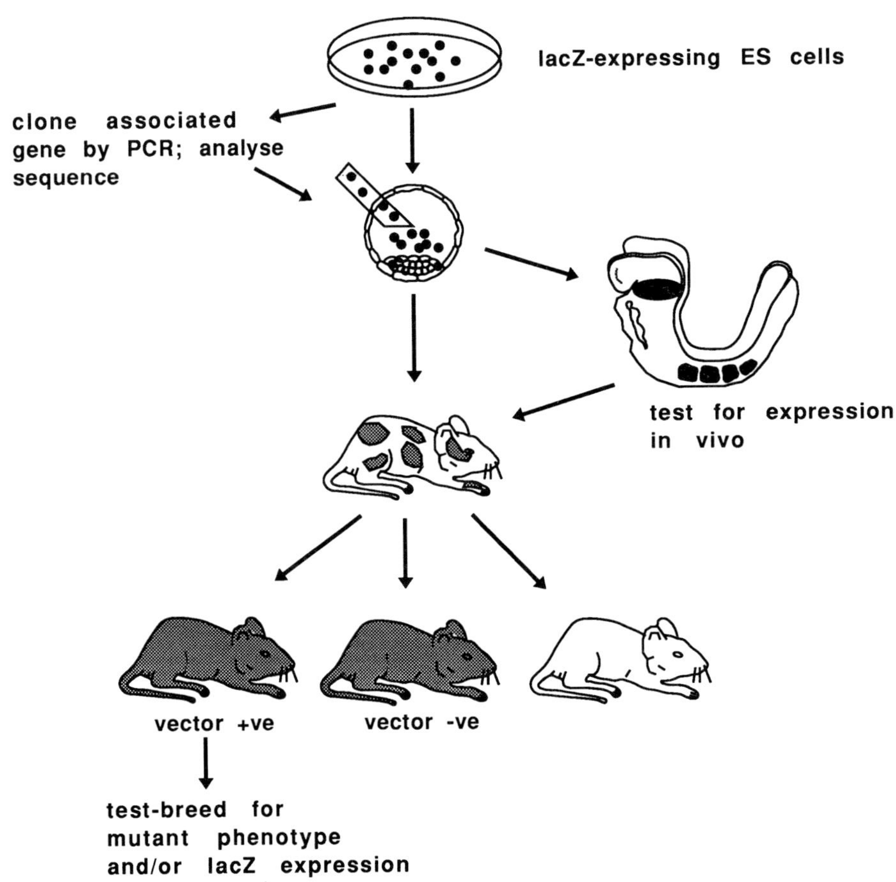

Figure 4. Strategies for large-scale gene-trap screens.

embryogenesis. Various strategies for large-scale screening can be envisaged, using different criteria to preselect for integrations of interest (figure 4). One could clone associated exons and use sequence information as the criterion for further study. Alternatively, expression in specific patterns during embryogenesis could be the method of choice. This is the approach we are using, taking advantage of the widespread contributions of ES cells after blastocyst injection to screen for patterns of expression of individual clones around gastrulation in chimeras. Of 230 cell lines examined to date, 60% showed expression at 8.5 days of development and 1/3 of these showed spatially restricted patterns (S.-L. Ang, A. Auerbach, S. Gasca, F. Guillemot, D. Hill, A. Joyner, J. Rossant and W. Wurst, unpublished data). Alternatively, populations of *lacZ*-expressing ES cells could be introduced into chimeras such that male chimeras could carry several different integrations. Offspring could be scored for expression pattern. Finally, mutant phenotype could be the primary criterion, although the numbers that could be screened in this way are much lower than for a primary expression screen.

A major limitation of the gene trap approach is that the low efficiency of *lacZ* activation makes it not very feasible to screen for genes that are not expressed in ES cells. Although many genes that are involved in early embryonic development may be expressed in ES cells, it is unlikely that *lacZ*-expressing gene trap ES cell lines will be a very useful resource for those interested in identifying and mutating genes involved in later organogenesis and terminal cell differentiation. To explore these areas, the *in vitro* differentiation capacity of ES cells will need to be exploited. ES cells have been shown to differentiate reproducibly into hematopoietic lineages (Wiles & Keller 1991), vascular endothelium (Wang *et al.* 1992) and muscle (Sanchez *et al.* 1991) *in vitro*. More effort needs to be placed on establishing conditions for reproducibly directing the differentiation of ES cells into different pathways, so that one could screen for gene trap integrations that became activated upon differentiation. Other screens using gene trap integrations to visualize genetic responses to particular growth and differentiation factors can also be envisaged.

4. THE FUTURE OF MOUSE GENOME MANIPULATION IN ES CELLS

The currently available technologies in ES cells have opened up many new possibilities in mammalian genetics. Mutations can be generated at will in cloned genes thought to be involved in any biological process, including those involved in human genetic disease. Novel insertional mutations that can be readily cloned can also be generated in large numbers. However, the full extent of genetic manipulation possible with ES cells has certainly not been reached. We can expect to see further refinement of targeting technology to increase targeting efficiencies and allow easy generation of site-specific mutations. Targeting mutations to specific tissues using site-specific recombinases may also be possible. More sophisticated entrapment vectors for random mutagenesis could be envisaged as well.

Methods that rely on the mouse as the final assay system will continue to be slow and expensive. The beauty of ES cells is that they provide the opportunity for establishing genetic screens *in vitro* in cell culture, where there is no limit on numbers of events that can be screened. This fact has not been fully exploited as yet, but one can, for example, envisage generation of sets of ES lines with specific chromosome deletions or truncations, for genomic analysis and phenotype assessment. It should also be possible to devise strategies to screen for dominant suppressors or enhancers *in vitro*. Combined with improvements in directing the differentiation of ES cells *in vitro*, these approaches promise to provide a wealth of genetic information on mammalian development and differentiation.

We thank A. Joyner for continued input and useful discussion. The authors' own work was supported by grants from the Medical Research Council and the National Cancer Institute of Canada, the National Institutes of Health (U.S.A.) and Bristol-Myers Squibb Ltd. J.R. is an International Scholar of the Howard Hughes Medical Institute and a Terry Fox Cancer Research Scientist of the NCIC. C.M. holds an MRC studentship.

REFERENCES

Beddington, R.S.P. & Robertson, E.J. 1989 An assessment of the developmental potential of embryonic stem cells in the midgestation mouse embryo. *Development* **105**, 733–737.

Bernelot-Moens, C., Auerbach, B.A., Conlon, R.A., Joyner, A.L. & Rossant, J. 1992 A targeted mutation reveals a role for N-*myc* in branching morphogenesis in the embryonic mouse lung. *Genes Dev.* **6**, 691–704.

Bradley, A., Evans, M. & Kaufman, M.H. 1984 Formation of germline chimaeras from embryo-derived teratocarcinoma cell lines. *Nature, Lond.* **309**, 255–256.

Brenner, D.G., Lin-Chao, S. & Cohen, S.N. 1989 Analysis of mammalian cell genetic regulation *in situ* by using retrovirus-derived "portable exons" carrying the *Escherichia coli LacZ* gene. *Proc. natn. Acad. Sci. U.S.A.* **86**, 5517–5521.

Boulter, C.A., Aguzzi, A., Williams, R.L., Wagner, E.F., Evans, M.J. & Beddington, R. 1991 Aberrant development and twinning in chimaeric mice. *Development* **111**, 357–366.

Capecchi, M.R. 1989 Altering the genome by homologous recombination. *Science, Wash.* **244**, 1288–1292.

Charron, J., Malynn, B.A., Robertson, E.J., Goff, S.P. & Alt, F.W. 1990 High-frequency disruption of the N-*myc* gene in embryonic stem and pre-B cell lines by homologous recombination. *Molec. Cell Biol.* **10**, 1799–1804.

Chen, U. 1992 Differentiation of mouse embryonic stem cells to lympho-hematopoietic lineages *in vitro*. *Devl. Immunol.* **2**, 29–50.

Conlon, F.L., Barth, K.S. & Robertson, E.J. 1991 A novel retrovirally induced embryonic lethal mutation in the mouse: Assessment of the developmental fate of embryonic stem cells homozygous for the 413.d proviral integration. *Development* **111**, 969–981.

Covarrubias, L., Nashida, Y., Terao, M., D'Eustachio, P. & Mintz, B. 1987 Cellular DNA rearrangements and early developmental arrest caused by DNA insertion in transgenic mouse embryos. *Molec. Cell Biol.* **7**, 2243–2247.

Dale, E.C. & Ow, D.W. 1991 Gene transfer with subsequent removal of the selection gene from the host genome. *Proc. natn. Acad. Sci. U.S.A.* **88**, 10558–10562.

Doetschman, T., Maeda, N. & Smithies, O. 1988 Targeted mutation of the Hprt gene in mouse embryonic stem cells. *Proc. natn. Acad. Sci. U.S.A.* **85**, 8583–8587.

Doetschman, T.G., Eistetter, M., Katz, M., Schmidt, W. & Kemler, R. 1985 *In vitro* development of blastocyst-derived embryonic stem cell lines: Formation of visceral yolk sac, blood islands and myocardium. *J. Embryol. exp. Morph.* **87**, 27–45.

Donehower, L.A., Harvey, M., Slagle, B.L., McArthur, M.J., Montgomery, C.A., Jr, Butel, J.S. *et al.* 1992 Mice deficient for p53 are developmentally normal but susceptible to spontaneous tumours. *Nature, Lond.* **356**, 215–222.

Evans, M. & Kaufman, M.H. 1981 Establishment in culture of pluripotential cells from mouse embryos. *Nature, Lond.* **292**, 154–155.

Friedrich, G. & Soriano, P. 1991 Promoter traps in embryonic stem cells: a genetic screen to identify and mutate developmental genes in mice. *Genes Dev.* **5**, 1513–1523.

Fung-Leung, W.-P., Schilham, A., Rahemtulla, A.M., Kundig, M., Vollenweider, J., Potter, J. *et al.* 1991 CD8 is needed for development of cytotoxic T cells but not helper T cells. *Cell* **65**, 443–449.

Gardner, R.L. 1968 Mouse chimaeras obtained by the injection of cells into the blastocyst. *Nature, Lond.* **220**, 596–597.

Golic, K.G. & Lindquist, S. 1989 The FLP recombinase of yeast catalyzes site-specific recombination in the *Drosophila* genome. *Cell* **59**, 499–509.

Gossler, A., Joyner, A.L., Rossant, J. & Skarnes, W.C. 1989 Mouse embryonic stem cells and reporter constructs to detect developmentally regulated genes. *Science, Wash.* **244**, 463–465.

Gridley, T., Soriano, P. & Jaenisch, R. 1987 Insertional mutagenesis in mice. *Trends Genet.* **3**, 162–166.

Gridley, T., Gray, D.A., Orr-Weaver, T., Soriano, P., Barton, D.E., Francke, U. *et al.* 1991 Molecular analysis of the *Mov 34* mutation: transcript disrupted by proviral integration in mice is conserved in *Drosophila*. *Development* **109**, 235–242.

Hasty, P., Ramirez-Solis, R., Kramlauf, R. & Bradley, A. 1991 Introduction of a subtle mutation into the *Hox-2.6* locus in embryonic stem cells. *Nature, Lond.* **350**, 243–246.

Jaenisch, R. 1976 Germ line integration and Mendelian transmission of the exogenous Moloney murine leukemia virus. *Proc. natn. Acad. Sci. U.S.A.* **73**, 1260–1264.

Jaenisch, R., Harbers, K., Schnieke, A., Lohler, J., Chumakov, I., Jahner, D. *et al.* 1983 Germline integration of Moloney murine leukemia virus at the *Mov 13* locus leads to recessive mutational and early embryonic death. *Cell* **32**, 209–216.

Jaenisch, R. 1988 Transgenic animals. *Science, Wash.* **240**, 1468–1674.

Johnson, R.S., Sheng, M., Greenberg, M.E., Kolodner, R.D., Papaioannou, V.E. & Spiegelman, B.M. 1989 Targeting of nonexpressed genes in embryonic stem cells via homologous recombination. *Science, Wash.* **245**, 1234–1236.

Jurgens, G., Wieschaus, E., Nusslein-Volhard, C. & Kluding, H. 1984 Mutations affecting the pattern of the larval cuticle in *Drosophila melanogaster II.* Zygotic loci on the third chromosome. *Wilhelm Roux's Arch.* **193**, 283–295.

Kadokawa, Y., Suemori, H. & Nakatsuji, N. 1990 Cell lineage analyses of epithelia and blood vessels in chimeric mouse embryos by use of an embryonic stem cell line expressing the β-galactosidase gene. *Cell Diff. Dev.* **29**, 187–194.

Kaufman, M.H. & Webb, S. 1990 Postimplantation development of tetraploid mouse embryos produced by electrofusion. *Development* **110**, 1121–1132.

Kerr, W.G., Nolan, G.P., Serafini, A.T. & Herzenberg, L.A. 1989 Transcriptionally defective retroviruses containing *lacZ* for the *in situ* detection of endogenous genes and developmentally regulated chromatin. *Cold Spring Harbor Symp. Quant. Biol.* **54**, 767–776.

Lang, Y., Bluethmann, H., Lipp, H.-P., DeArmond, S.J., Prusiner, S.B., Aguet, M. *et al.* 1992 Normal development and behaviour of mice lacking the neuronal cell-surface PrP protein. *Nature, Lond.* **356**, 577–582.

Lee, J.J., Radice, G., Perkins, C.P. & Costantini, F. 1992 Identification and characterization of a novel, evolutionarily conserved gene disrupted by the murine Hβ58 embryonic lethal transgene insertion. *Development* **115**, 277–288.

Lindenbaum, M.H. & Grosveld, F. 1990 An *in vitro* globin gene switching model based on differentiated embryonic stem cells. *Genes Dev.* **4**, 2075–2085.

Maas, R.L., Zeller, R., Woychik, R.P., Vogt, T.F. & Leder, P. 1990 Disruption of formin-encoding transcripts in two mutant *limb deformity* alleles. *Nature, Lond.* **346**, 853–855.

Mansour, S.L., Thomas, K.R. & Capecchi, M.R. 1988 Disruption of the proto-oncogene *int*-2 in mouse embryo-derived stem cells: a general strategy for targeting mutations to non-selectable genes. *Nature, Lond.* **336**, 348–352.

Martin, G.R. 1981 Isolation of a pluripotent cell line from early mouse embryos cultured in medium conditioned by teratocarcinoma stem cells. *Proc. natn. Acad. Sci. U.S.A.* **78**, 7634–7638.

McMahon, A.P. & Bradley, A. 1990 The *Wnt-1* (*int-1*) proto-oncogene is required for development of a large region of the mouse brain. *Cell* **62**, 1073–1085.

Mombaerts, P., Iacomini, J., Johnson, R.S., Herrup, K., Tonegawa, S. & Papaioannou, V.E. 1992 RAG-1-deficient mice have no mature B and T lymphocytes. *Cell* **68**, 869–878.

Mugrauer, G., Alt, F.W. & Ekblom, P. 1988 N-*myc* proto-oncogene expression during organogenesis in the developing mouse as revealed by *in situ* hybridization. *J. Cell Biol.* **107**, 1325–1335.

Nagy, A., Gocza, E., Diaz, E.M., Prideaux, V.R., Ivanyi, E., Markkula, M. *et al.* 1990 Embryonic stem cells alone are able to support fetal development in the mouse. *Development* **110**, 815–822.

O'Gorman, S., Fox, D.T. & Wahl, G.M. 1991 Recombinase-mediated gene activation and site-specific integration in mammalian cells. *Science, Wash.* **251**, 1351–1355.

Reddy, S., DeGregori, J.V., von Melchner, H. & Ruley, H.E. 1991 Retrovirus promoter-trap vector to induce *lacZ* gene fusions in mammalian cells. *J. Virol.* **65**, 1507–1515.

Reith, A.D., Rottapel, R., Giddens, E., Brady, C., Forrester, L. & Bernstein, A. 1990 *W* mutant mice with mild or severe developmental defects contain distinct point mutations in the kinase domain of the c-*kit* receptor. *Genes Dev.* **4**, 390–400.

Robertson, E.J., Bradley, A., Kuehn, M. & Evans, M. 1986 Germ-line transmission of genes introduced into cultured pluripotential cells by retroviral vector. *Nature, Lond.* **323**, 445–447.

Rossant, J., Merentes-Diaz, E., Gocza, E., Ivanyi, E. & Nagy, A. 1992 Developmental potential of mouse embryonic stem cells. In *Serono Symposium on Preimplantation*

Embryo Development (ed. B. Bavister). Springer Verlag. (In the press.)

Sanchez, A., Jones, W.K., Gulick, J., Doetschman, T. & Robbins, J. 1991 Myosin heavy chain gene expression in mouse embryoid bodies. An *in vitro* developmental study. *J. biol. Chem.* **266**, 22419–22426.

Sawai, S., Shimono, A., Hanaoka, K. & Kondoh, H. 1991 Embryonic lethality resulting from disruption of both N-myc alleles in mouse zygotes. *New Biol.* **3**, 861–869.

Shinkai, Y., Rathbun, G., Lam, K.-P., Ottz, E.M., Stewart, V., Mendelsohn, M. *et al.* 1992 RAG-2-deficient mice lack mature lymphocytes owing to inability to initiate V(D)J rearangement. *Cell* **68**, 855–868.

Skarnes, W.C., Auerbach, B.A. & Joyner, A.L. 1992 A gene trap approach in mouse embryonic stem cells: the *lacZ* reporter is activated by splicing, reflects endogenous gene expression and is mutagenic. *Genes Dev.* **6**, 903–918.

Smith, A.G., Heath, J.K., Donaldson, D.D., Wong, G.G., Moreau, J., Stahl, M. *et al.* 1988 Inhibition of pluripotential embryonic stem cell differentiation by purified peptides. *Nature, Lond.* **336**, 688–690.

Smithies, O., Gregg, R.G., Boggs, S.S., Kolaweski, M.A. & Kucherlapati, R. 1985 Insertion of DNA sequences into the human chromosome β-globin locus by homologous recombination. *Nature, Lond.* **317**, 230–234.

Snow, M.H.L. 1975 Embryonic development of tetraploid mice during the second half of gestation. *J. Embryol. Exp. Morph.* **34**, 707–721.

Stanton, B.R., Reid, S.W. & Parada, L.F. 1990 Germ line transmission of an inactive N-myc allele generated by homologous recombination in mouse embryonic stem cells. *Molec. Cell Biol.* **10**, 6755–6758.

Stewart, T.A. & Mintz, B. 1981 Successive generations of mice produced from an established culture line of emploid teratocarcinoma cells. *Proc. natn. Acad. Sci. U.S.A.* **78**, 6314–6318.

Tarkowski, A.K., Witkowska, A. & Opas, J. 1978 Development of cytochalasin β-induced tetraploid and diploid-tetraploid mosaic mouse embryos. *J. Embryol. exp. Morph.* **41**, 47–64.

Te Riele, H., Maandag, E.R., Clarke, A., Hooper, M. & Berns, A. 1990 Consecutive inactivation of both alleles of the *pim-1* proto-oncogene by homologous recombination. *Nature, Lond.* **348**, 649–651.

Te Riele, H., Robanus Maandag, E. & Berns, A. 1992 Highly efficient gene targeting in embryonic stem cells through homologous recombination with isogenic DNA constructs. *Proc. natn. Acad. Sci. U.S.A.* **89**, 5128–5132.

Thomas, K.R., Folger, K.R. & Capecchi, M.R. 1986 High frequency targeting of genes to specific sites in the mammalian genome. *Cell* **44**, 419–428.

Thomas, K.R. & Capecchi, M.R. 1987 Site-directed mutagenesis by gene targeting in mouse embryo-derived stem cells. *Cell* **51**, 503–512.

Thomas, K.R. & Capecchi, M.R. 1990 Targeted disruption of the murine *int-1* proto-oncogene resulting in severe abnormalities in midbrain and cerebellar development. *Nature, Lond.* **346**, 847–850.

Thompson, S., Clarke, A.R., Pow, A.M., Hooper, M.L. & Melton, D.W. 1989 Germ line transmission and expression of a corrected HPRT gene produced by gene targeting in embryonic stem cells. *Cell* **56**, 313–321.

Valancius, V. & Smithies, O. 1991 Testing an "in-out" targeting procedure for making subtle genomic modifications in mouse embryonic stem cells. *Molec. Cell Biol.* **11**, 1402–1408.

von Melchner, H., DeGregori, J.V., Rayburn, H., Reddy, S., Friedel, C. & Ruley, H.E. 1992 Selective disruption of genes expressed in totipotent embryonal stem cells. *Genes Dev.* **6**, 919–927.

Wang, R., Clark, R. & Bautch, V.L. 1992 Embryonic stem cell-derived cystic embryoid bodies form vascular channels—an *in vitro* model of blood vessel development. *Development* **114**, 303–316.

Weiher, H., Noda, T., Gray, D.A., Sharpe, A.H. & Jaenisch, R. 1990 Transgenic mouse model of kidney disease: Insertional inactivation of ubiquitously expressed gene leads to nephrotic syndrome. *Cell* **62**, 425–434.

Wieschaus, E., Nusslein-Volhard, C. & Jurgens, G. 1984 Mutations affecting the pattern of the larval cuticle in *Drosophila melanogaster III.* Zygotic loci on the X-chromosome and fourth chromosome. *Wilhelm Roux's Arch.* **193**, 296–307.

Wiles, M.V. & Keller, G. 1991 Multiple hematopoietic lineages develop from embryonic stem (ES) cells in culture. *Development* **111**, 259–267.

Wilkie, T.M. & Palmiter, R.D. 1987 Analysis of the integrant of Myk-103 transgenic mice in which males fail to transmit the integrant. *Molec. Cell Biol.* **7**, 1646–1655.

Williams, R.L., Hilton, D.J., Pease, S., Wilson, T.A., Stewart, C.L., Gearing, D.P. *et al.* 1988 Myeloid leukaemia inhibitory factor maintains the development potential of embryonic stem cells. *Nature, Lond.* **336**, 684–687.

Zimmerman, K., Yancopoulos, G.D., Collum, R.G., Smith, R.K., Kohl, N.E., Denis, K.A. *et al.* 1986 Differential expression of myc family genes during murine development. *Nature, Lond.* **319**, 780–783.

10

Gene transfer therapy for heritable disease: cell and expression targeting

K. Mitani, P. R. Clemens, A. B. Moseley and C. T. Caskey

SUMMARY

Gene therapy is defined as the delivery of a functional gene for expression in somatic tissues with the intent to cure a disease. Different gene transfer strategies may be required to target different tissues. Adenosine deaminase (ADA) deficiency is a good gene therapy model for targeting a rare population of pluripotent hematopoietic stem cells capable of self-renewal. We present evidence for the highly efficient gene transfer and sustained expression of human ADA in human primitive hematopoietic progenitors using retroviral supernatant with a supportive stromal layer. A stem cell-enriched ($CD34^+$) fraction was also successfully transduced.

Duchenne muscular dystrophy (DMD) is also a good model for somatic gene therapy. Two of the challenges presented by this model are the large size of the gene and the large number of target cells. Germline gene transfer and correction of the phenotype has been demonstrated in transgenic *mdx* mice using both a full-length and a truncated form of the dystrophin cDNA. We present here a deletion mutagenesis strategy to truncate the dystrophin cDNA such that it can be accommodated by retroviral and adenoviral vectors useful for somatic gene therapy.

1. INTRODUCTION

To date, clinical therapy for most genetic diseases has been limited to supportive care and, when an enzyme defect has been identified, protein replacement. Recent developments in the understanding of the molecular basis of numerous genetic diseases has facilitated novel experimental approaches to therapy. We present advances directed toward clinically effective gene transfer therapy for two single-gene disorders, adenosine deaminase (ADA) deficiency and Duchenne muscular dystrophy (DMD). Each provides quite different challenges.

Hematopoietic stem cells can be purified, grown *in vitro*, and will repopulate and reconstitute the human hematopoietic system when administered intravenously. ADA deficiency, which results in a severe combined immune deficiency, therefore serves as a model for gene therapy with a high likelihood of success. In contrast, DMD results from a structural gene defect in muscle cells which comprise 75% of the human body. Myoblasts (muscle precursor cells) lack migration and targeting properties, and gene therapy of DMD is therefore likely to be achieved by organ gene delivery *in vivo*, rather than cellular delivery or gene delivery *in vitro*.

2. BACKGROUND

(*a*) *Adenosine deaminase deficiency*

ADA deficiency, an autosomal recessive disorder, accounts for 25% of severe combined immunodeficiency (SCID). This metabolic enzyme catalyses the irreversible deamination of adenosine and deoxyadenosine (dAdo) to inosine and deoxyinosine, respectively. In the absence of ADA, dAdo accumulates in many tissues and the accumulation of dAdo and its metabolites, particularly deoxyadenosine triphosphate, inhibits DNA synthesis, resulting in profound T-cell and subsequent B-cell dysfunction. Approximately 100 families with ADA-deficient SCID have been identified. Untreated, ADA-deficient children usually die before age 2 from overwhelming infection (Kredich *et al.* 1989).

Complementary DNA (cDNA) sequences of normal human ADA have been cloned and characterized (Daddona *et al.* 1984) allowing analysis of seven mutant ADA alleles from five ADA-deficient individuals (Kredich *et al.* 1989). Five of these alleles have a single base mutation resulting in amino acid substitutions that presumably decrease ADA activity. The absence of enzyme activity in a hemolysate of the patientsa erythrocytes is diagnostic for this disease. Assay of ADA in cultured amniotic cells, fetal blood, and chorionic villi allows prenatal diagnosis and carrier detection (Kredich *et al.* 1989).

(*b*) *Duchenne muscular dystrophy*

DMD, an X-linked recessive disorder affecting skeletal and cardiac muscle and brain, is the most common form of muscular dystrophy in children affecting 1 in 3500 male births. DMD and its milder allelic form, Becker muscular dystrophy (BMD), are caused by mutations in the dystrophin gene, which codes for a cytoskeletal structural protein. DMD is a degenerative disorder that begins with skeletal muscle dysfunction early in life, progressing to loss of ambulation, respiratory compromise, cardiac dysfunction and death by the second or third decade of life. A proportion of DMD patients also exhibit mental deficiency that cannot be entirely explained as secondary effects of social environment and functional disability. The phenotype of BMD presents later than DMD and has a more gradual and variable degenerative course (Clemens & Caskey 1992).

The cloning of the cDNA for the dystrophin gene (Koenig *et al.* 1987) revolutionized DNA diagnostic strategies for DMD. Use of dystrophin cDNA probes for Southern hybridization demonstrates large deletions or duplications in approximately 70% of DMD patients (Darras *et al.* 1988; Baumbach *et al.* 1989; Gilgenkrantz *et al.* 1989; Hu *et al.* 1989; Koenig *et al.* 1989; Angelini *et al.* 1990; Hu *et al.* 1990). Approximately 98% of the deletions can also be detected by the simpler technique of polymerase chain reaction (PCR) exon amplification (Chamberlain *et al.* 1988, 1989; Beggs *et al.* 1990; Multi Center Study Group 1992). The remaining 30% of DMD cases are probably due to mutations below the detection threshold of Southern and PCR studies. There are two reports of DMD patients with point mutations causing premature chain termination within the dystrophin gene. These mutations were detected by prediction of the mutation site based on a truncated protein fragment observed by Western analysis, followed by DNA sequencing (Bulman *et al.* 1991) and by identification of a new *Hin*dIII site within exon 48 followed by DNA sequencing (Clemens *et al.* 1992). Strategies to screen for point mutations within the dystrophin gene are also being developed (Kilimann *et al.* 1992; Roberts *et al.* 1992). Despite the remarkable advances in DNA diagnostics for DMD and the application of these strategies to prenatal diagnosis (Ward *et al.* 1989; Clemens *et al.* 1991), a high spontaneous mutation rate within the gene will perpetuate this devastating disorder.

3. CURRENT OPTIONS FOR THERAPY

(*a*) *Adenosine deaminase deficiency*

Bone marrow transplantation (BMT) from a human lymphocyte antigen (HLA)-identical donor is the preferred treatment for ADA deficiency, and while associated with morbidity, can result in complete cure. Unfortunately, fewer than 30% of the patients have an HLA matched sibling. Enzyme replacement is an attractive mode of therapy because dAdo is freely diffusible, and therefore the ADA enzyme does not need to enter the target cells to be partially effective. Weekly injections of bovine ADA protein conjugated to polyethylene glycol (PEG-ADA) have been shown to reverse the biochemical abnormalities, and restore some T-lymphocyte function. Neverthless, total immune reconstitution has not been demonstrated and anti-PEG-ADA antibodies have been demonstrated in some patients(Kredich *et al.* 1989).

(*b*) *Duchenne muscular dystrophy*

To date, the only treatment available to DMD patients is symptomatic, comprising physical therapy, tendon release to relieve joint contracture, and spine surgery for scoliosis (Brooke *et al.* 1989). These procedures may improve the quality of life, but do not prevent the inevitable degeneration and death. Recent studies exploring immunosuppression with prednisone show promise for slowing the course of muscle degeneration, but not for eradication of the disease (Mendell *et al.* 1989; Fenichel *et al.* 1991; Griggs *et al.* 1991). There is a clear need for more effective treatment for patients with DMD. Two strategies currently under study as potential somatic therapies for the disease are myoblast transfer and gene transfer (discussed below).

4. ADVANCES TOWARD GENE THERAPY

Gene therapy is defined as the transfer of a functional gene into the cells of an organism for the treatment of disease (Miller 1990; Levine *et al.* 1991; Mulligan 1991; Verma *et al.* 1991). In general, a disease that might be amenable to gene therapy should be a single-gene disorder of recessive inheritance and be well understood at the molecular level. The biology of the target tissue should be known and gene transfer into the target tissue should result in levels of expression adequate to cure the disease. ADA deficiency is an ideal model for gene therapy targeting hematopoietic stem cells (HSCs) based on the results from BMT, suggesting that transfer of the normal ADA gene into HSCs would be sufficient to correct the enzymatic defect. Only 20% of the normal ADA level in the blood would be needed for treatment and it is plausible that *in vivo* selection of even a few corrected stem cells would take place. In DMD, the defects are well understood at the molecular level. However, further understanding of the function of dystrophin in tissues is necessary for successful gene transfer strategies. *Ex vivo* gene transfer into candidate target cells, myoblasts, has been attempted. The *mdx* mouse provides an animal model for DMD gene therapy.

Highly efficient gene transfer is required for gene therapy. The retroviral vector is the best characterized viral vector for human gene transfer (Levine *et al.* 1991; Mulligan 1991). Retroviral vectors transduce dividing cells in culture with almost 100% efficiency

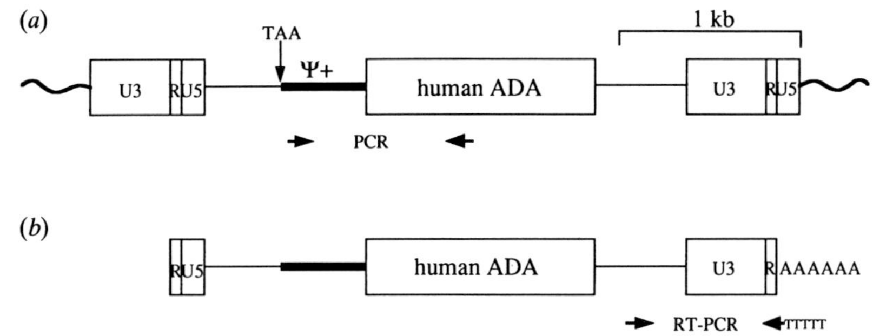

Figure 1. Structure of integrated retroviral vector (*a*) and transcript from the vector (*b*). In this vector, ΔN2stADA, the start codon of gag$^+$ was mutated to a stop codon (TAA). Human ADA cDNA is transcribed from the 5′ LTR. The primers for the provirus PCR (*a*) and vector mRNA RT-PCR (*b*) are indicated as arrows. Wavy lines indicate flanking chromosomal DNA.

and integrate into the host genome stably without causing gross rearrangements.

(*a*) *In adenosine deaminase deficiency*

A gene replacement therapy trial in two patients with ADA deficiency began recently at the National Institutes of Health (Culver *et al.* 1991). The ADA gene transfer strategy targets peripheral T-cells that have been stimulated *in vitro*. Both patients have shown improvement clinically, as well as in *in vitro* and *in vivo* immune function. However, because only mature T-cells are transduced, not all of the immune function indices have normalized (Anderson 1992). Also, the limited life span of T-cells necessitates repetitive infusions of transduced T-cells, raising serious concerns about clonality of the immune cells and the possibility of insertional mutagenesis of cancer-related genes by repeated retroviral transfections.

The HSC, which constitutes 0.01–0.1% of human bone marrow (BM) cells, has been regarded as an ideal target tissue for human gene therapy because it is pluripotent and undergoes self-renewal. In theory, all hematopoietic cell lineages would demonstrate the transduced gene after the transfection of even a few HSCs. Cells bearing the CD34 antigen represent a population of bone marrow cells thought to contain the most primitive progenitors. Retroviral vectors efficiently transfer genes into mouse HSCs; we have also used them for gene targeting to human HSCs.

A retroviral vector, ΔN2stADA (figure 1) (Scarpa *et al.* 1991), which utilizes the retroviral LTR to drive human ADA, was used for transduction studies. Low density mononuclear cells from normal BM samples were transduced either by co-cultivation with the virus producing cell line or by virus-containing supernatant in the presence of a supportive autologous stromal layer. Transduced BM cells were kept in myeloid long term culture (LTC) and a colony-forming assay on methylcellulose-containing medium was performed as described previously (Cournoyer *et al.* 1991). Polybrene, recombinant human interleukin 3 (rhIL-3) and rhIL-6 (gifts from Dr S. C. Clark, Genetics Institute) were added during transduction in both protocols. Additionally, CD34$^+$ cells were selected on anti-CD34 selection devices (Applied Immune Sciences), plated onto previously grown autologous stroma and infected with retroviral supernatant. The cells were maintained in LTC before analysis.

PCR and reverse transcriptase-PCR (RT-PCR) were used to examine the integration and expression, respectively, of ADA-retrovirus vector in human hematopoietic colonies derived from LTC (figure 1). Human clonogenic progenitors, which are present at week 0, disappear by death or differentiation after five weeks in myeloid LTC. Therefore, colonies formed after that time are derived from primitive progenitors, which are presumed to be closely related or identical to HSCs present at the time of transduction (Sutherland *et al.* 1989; Sutherland *et al.*, 1990). Granulocyte–macrophage (GFU-GMS) colonies demonstrated approximately 90% transduction at week 0, 50% transduction at week 3 and 50% later than week 5, indicating that approximately 50% of primitive hematopoietic progenitors were transduced by the co-cultivation method (figure 2*a*). The percentage of transcript-positive colonies was 33% at week 0, 22% at week 3 and 14% at week 6 (figure 2*a*). At week 6, the transcript was also detected in 25% of colony-forming unit-erythroids (GFU-ES) (data not shown). The percentage of provirus-positive and transcript-positive colonies transduced with viral supernatant was equivalent to that of co-cultivation (figure 2*b*). While transduction with viral supernatant is preferred to co-cultivation, previous results with viral supernatant have demonstrated poor transfer efficiency (Eglitis *et al.* 1988; Bodine *et al.* 1991; Cournoyer *et al.* 1991). The stromal support appears to play a major role in efficient gene transfer by viral supernatant. The CD34$^+$ fractions from normal individuals, as well as from an ADA-deficient patient, were also transduced successfully, suggesting that the viral supernatant protocol is applicable to a clinical trial in ADA-deficient patients (data not shown).

To detect the increased ADA activity after retroviral gene transfer, an ADA conversion assay was performed on collected non-adherent cells in LTC (Aitken *et al.* 1980). The activity of purine nucleoside phosphorylase (PNP) was used to standardize the amount of protein in each samples. We observed a 3.6-fold increase of ADA to PNP activity at week 5 post-transduction that was maintained at week 9 (figure 3). In the case of ADA(−) BM, the ratio of

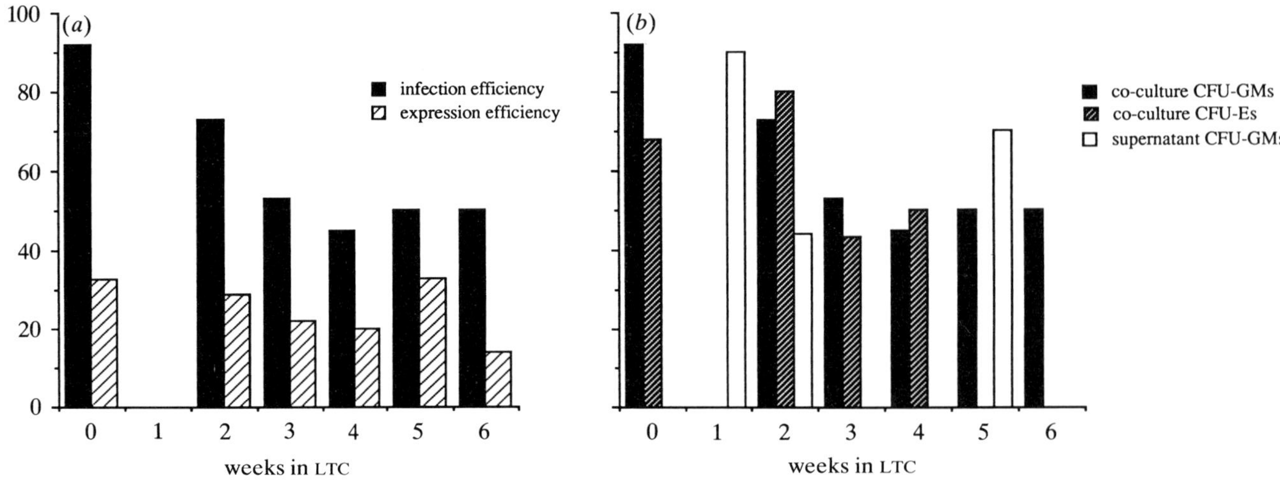

Figure 2. (*a*) Gene transfer and expression efficiency of the retroviral vector in GFU-GMS transduced by co-cultivation. (*b*) Comparison of gene transfer efficiency by co-cultivation protocol and viral supernatant protocol.

ADA to PNP activity was 21.0 in the transduced BM and 0.22 in untransduced BM ten weeks post-transduction (data not shown). Although there could be a small amount of contaminating long-lived mature cells in this fraction, the large increase of ADA activity at this late stage of LTC suggests strong expression of the transduced ADA gene in cells differentiated from transduced progenitors. Development of a mouse model of human ADA deficiency, in progress in our laboratory, is also important to evaluate the efficacy of our vector. The expression from the LTR promoter of Moloney murine leukemia virus (Mo-MuLV) is sustained in both clonogenic progenitors and mature cells derived from transduced primitive progenitors *in vitro*. Transplantation of transduced BM into *Bg/nu/xid* mice (Dick *et al.* 1991) or transduced BM depleted of mature T- and B-lymphocytes and BM from mice with severe combined immunodeficiency into lethally irradiated Balb/c mice (Lubin *et al.* 1991) would be ideal for evaluation of ADA gene replacement into lymphocytes. *Bg/nu/xid* mice were used to evaluate the efficacy of ADA gene transfer into peripheral T-cells of an ADA deficient patient (Ferrari *et al.* 1991). Long-term survival of vector-transduced human lymphocytes as well as restored immune functions, as indicated by the presence of human immunoglobulin and antigen-specific T-cells, was demonstrated in recipient animals (Ferrari *et al.* 1991).

(*b*) *Duchenne muscular dystrophy*

The cloning of the full-length dystrophin cDNA

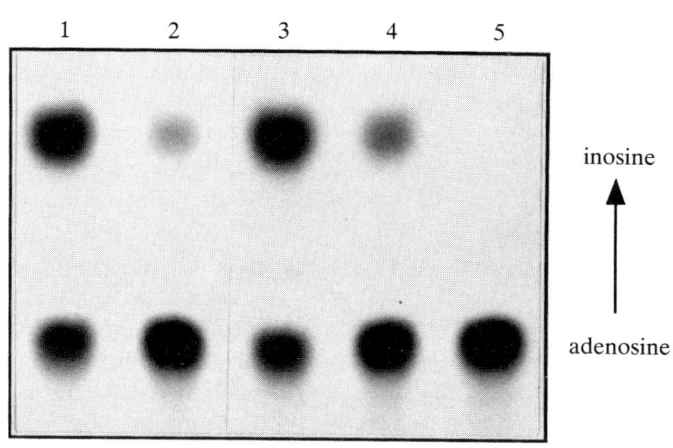

	1	2	3	4	5
ADA activity	1.27	0.12	1.30	0.26	---
PNP activity	0.67	0.27	0.19	0.30	---
ADA/PNP	1.82	0.51	2.90	0.79	---
increase	3.6		3.7		

Figure 3. Proviral induced ADA activity in normal BM. A ratio of ADA to PNP activity (ADA/PNP) is an average of duplicated assays. ADA and PNP activities are described as (nanomole conversion) per hour per 10^3 cells. Lanes : 1 and 2, 5 weeks post-transduction; 3 and 4, 9 weeks post-transduction; 5, no extract; 1 and 3, transduced BM; 2 and 4, untransduced BM.

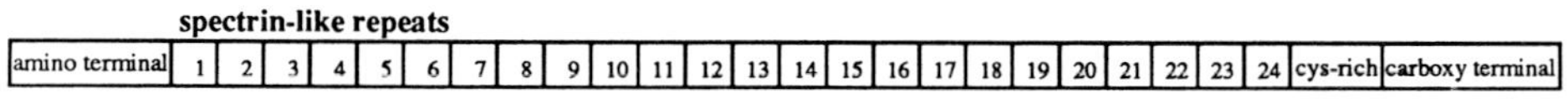

Figure 4. Schematic representation of the structure of dystrophin predicted from the cDNA sequence.

that was shown to express dystrophin in cell culture (Acsadi *et al.* 1991; Dickson *et al.* 1991; Lee *et al.* 1991a) has facilitated two general experimental approaches toward gene replacement therapy for DMD: (i) demonstration of germline correction in transgenic mice; and (ii) studies toward somatic cell therapy, both by myoblast transfer and gene transfer.

Germline correction of dystrophin deficiency in animal models provides an experimental demonstration of the potential for gene replacement therapy for DMD patients. A transgenic mouse with a single copy integration of the full-length dystrophin cDNA driven by the muscle creatine kinase promoter with the E1 and E2 enhancers was bred onto the dystrophin-deficient *mdx* genetic background. Transgene expression of both messenger RNA and protein was found in the expected pattern of tissues (Lee *et al.* 1991*b*). Another transgenic mouse was generated using a deleted dystrophin cDNA construct derived from a BMD patient and driven by the constitutive promoter elements of the Mo-MuLV (Wells *et al.* 1992).

Studies aimed at somatic therapy for DMD patients began with myoblast transfer. This is a form of cellular somatic therapy designed to exploit the normal biology of skeletal muscle regeneration to correct a genetic defect. Precursor mononuclear muscle cells (commonly called myoblasts or satellite cells) fuse with other myoblasts and existing muscle fibres to produce the multinucleate syncytia of differentiated skeletal muscle (Grounds 1991). The concept underlying myoblast transfer therapy for DMD is that normal donor myoblasts injected into degenerating, dystrophin-deficient muscle will fuse with the degenerating muscle fibres, provide dystrophin which is presumed to be necessary to maintain cytoskeletal integrity, and aid in the regeneration of skeletal muscle. Promising results were observed in animal studies in which normal mouse donor myoblasts, injected into *mdx* mouse muscle, fused with host dystrophin-deficient fibres forming mosaic myofibres (Law *et al.* 1988; Karpati *et al.* 1989*a,b*; Partridge *et al.* 1989; Morgan *et al.* 1990). Pilot human myoblast transfer studies, in which myoblasts derived from a muscle biopsy taken from a related donor are injected into selected, small muscles of patients with DMD, are now in progress (Law *et al.* 1990, 1991; Huard *et al.* 1991; Miller *et al.* 1992). To date, myoblast transfer studies in humans have not yielded promising results, perhaps due in part to immune rejection of the donor myoblasts by HLA class I and class II specific alloreactive T cells and natural killer cells (Karpati *et al.* 1988; Bao *et al.* 1990; Hohlfeld & Engel 1990*a,b*; Mantegazza *et al.* 1991). Furthermore, a major limitation of myoblast transfer as a potential therapy for DMD is that it can only address the skeletal muscle manifestations of the disease. In addition, DMD patients have widespread skeletal muscle degeneration in many muscle groups and transplanted myoblasts may not migrate far from the injection site.

Other researchers have attempted to introduce the deficient genetic material more directly. One experimental attempt of somatic gene therapy for DMD *in vivo* was the intramuscular injection of human dystrophin DNA constructs into *mdx* mice resulting in low-level human dystrophin expression (Acsadi *et al.* 1991). These results are encouraging for the prospect of human gene therapy in DMD patients because they demonstrate that myoblasts or muscle fibres will incorporate and express exogenous DNA *in vivo*. Another means of direct gene transfer under investigation utilizes particle bombardment technology to introduce marker genetic material into cultured muscle cells and muscle tissue *in vivo* (Yang *et al.* 1990). To date, neither of these direct gene transfer strategies into muscle have achieved the efficiency of gene transfer that would be required for clinical efficacy.

The use of retroviral and adenoviral vectors holds promise as an efficient method of somatic gene transfer for DMD patients. The successful application of retroviral-mediated gene transfer into myoblasts has been demonstrated by transfer of the human growth hormone gene into mouse myoblasts (Barr & Leiden 1991; Dhawan *et al.* 1991) and by transfer of the human multidrug transporter gene into rat myoblasts (Salminen *et al.* 1991). Furthermore, retrovirus-infected myoblasts have been injected into recipient animal muscle resulting in expression of the gene insert that was sustained for one to three months (Barr & Leiden 1991; Dhawan *et al.* 1991; Salminen *et al.* 1991). An adenoviral vector containing the β-galactosidase reporter gene under the control of muscle-specific regulatory sequences has been used to demonstrate efficient transfection of myotubes in culture and of muscle after direct injection *in vivo* (Quantin *et al.* 1992).

Several unique challenges are presented by the application of retroviral and adenoviral vectors to gene transfer for DMD patients. First, the dystrophin cDNA is approximately twice the maximum insert size of both vectors. Second, the protein product of the transferred gene must be targeted to tissues that normally express dystrophin, and within muscle cells must localize to the cell membrane to perform its presumed cytostructural function.

Four domains of protein structure have been predicted from the DMD cDNA sequence: (i) an amino terminal domain; (ii) a repeat domain predicted to have a rod-like structure; (iii) a cysteine-rich domain; and (iv) a carboxy terminal domain (figure 4). The first three domains have homology to two known cytoskeletal proteins, α-actinin (domains one and three) and spectrin (domain two) (Koenig *et al.* 1988).

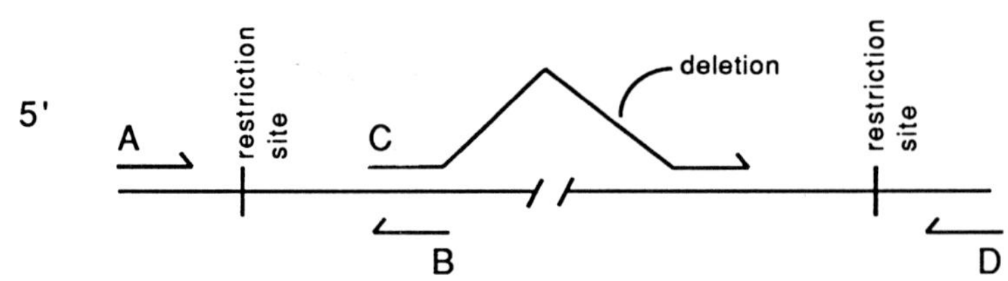

Figure 5. Diagram of the PCR-based deletion mutagenesis strategy. Primer C is a central primer that encodes the deletion such that the 5′ half of the primer is complementary to the dystrophin cDNA sequence just 5′ to the deletion and the 3· half of the primer is complementary to the dystrophin cDNA sequence just 3′ to the deletion. Primer B is complementary to the 5′ half of primer C. Primers A and D are flanking primers that are 5′ and 3′, respectively, of restriction sites used for cloning.

The fourth domain has homology to the autosomal transcript, DMDL (Love *et al.* 1989), which codes for dystrophin-related protein.

To truncate the dystrophin cDNA such that it can be accommodated by retroviral and adenoviral vectors, we are generating recombinant dystrophin cDNA constructs with deletions of various sizes in the spectrin-like repeat region of dystrophin. This region is targeted for deletion mutagenesis because the patients with milder phenotypes generally have in-frame deletions in this region (Gospe *et al.* 1989; England *et al.* 1990). The use of a PCR-based deletion mutagenesis strategy provides control over the exact position of the deletion and therefore, maintenance of the reading frame. Four oligonucleotide primers are designed for each engineered deletion, with the central primer (primer C) encoding the desired deletion (figure 5). PCR is performed with primers A and B and with primers C and D in two separate reactions. Products of the two reactions are mixed and a second PCR is performed with primers A and D. Under the conditions of the PCR, the products of the first PCR reactions denature allowing for the annealing of the strand complementary to primer A with the strand complementary to primer D followed by 3′ extension of each strand prior to amplification by the second PCR. After sequencing the PCR product to ascertain that no errors are introduced by the PCR, the fragment encoding the deletion is cloned into the original 5′ clone of the dystrophin cDNA. Following this, the mutagenized 5′ clone is spliced to the 3′ clone of the dystrophin cDNA to reconstitute a full-length recombinant dystrophin cDNA with a central deletion of part of the spectrin-like repeat region.

Using this approach, we have generated two truncated dystrophin constructs deleting 3–5 kb of the spectrin-like repeat region of dystrophin. By also truncating the 3′ untranslated region, the size of the DNA molecule can be reduced further. The truncated cDNA molecules will be tested for expression and targeting *in vitro* and *in vivo*.

5. CONCLUSIONS

Human gene therapy has progressed from theory to reality in a short time. The first gene therapy protocol, for correction of ADA deficiency in peripheral blood lymphocytes, began in September 1990. Now there are more than 20 active or approved clinical protocols and many additional protocols are in various stages of development. Several years ago, targeting of hematopoietic stem cells by retroviral vectors was thought to be the only realistic protocol for human gene therapy. Now, work is focused on a variety of somatic tissues, including muscle cells, and several gene transfer strategies.

The continued characterization of genes that are mutated in single gene disorders will widen the field of inherited diseases that can be approached by human gene therapy. Emphasis must be placed on the development of vectors or other transfer strategies that can deliver and express foreign genes more efficiently. Finally, it is likely that *in vivo* gene transfer will be required in some cases and preferable to *ex vivo* gene transfer, where cells are removed from a patient, the gene is transferred *in vitro* and the corrected cells are transplanted to the patient.

We thank Dr A. B. Deisseroth and Dr C. L. Reading from University of Texas M.D. Anderson Cancer Center for bone marrow samples from transplantation donors, Dr R. H. Buckley and Dr M. L. Market from Duke University for a bone marrow sample from an ADA-deficient patient, and Dr S. C. Clark for recombinant hematopoietic growth factors. This work was supported by USPHS grants from the National Institute of Health R01 DK42696 (C.T.C.), a Task Force on Genetics Grant from the Muscular Dystrophy Association and the Howard Hughes Medical Institute. C.T.C. is an Investigator of the Howard Hughes Medical Institute.

REFERENCES

Acsadi, G., Dickson, G., Love, D.R. *et al.* 1991 Human dystrophin expression in mdx mice after intramuscular injection of DNA constructs. *Nature, Lond.* **352**, 815–818.

Aitken, D.A., Kleijer, W.J., Niermeijer, M.F., Herb-Schleb-Voogt, E. & Galjaard, H. 1980 Prenatal detection of a probable heterozygote for ADA deficiency and severe combined immunodeficiency disease using a microradioassay. *Clin. Genet.* **17**, 293–298.

Anderson, W.F. 1992 Human gene therapy. *Science, Wash.* **256**, 808–813.

Angelini, C., Beggs, A.H., Hoffman, E.P., Fanin, M. & Kunkel, L.M. 1990 Enormous dystrophin in a patient with Becker muscular dystrophy. *Neurology* **40**, 808–812.

Bao, S.S., King, N.J. & dos Remedios, C.G. 1990 Elevated MHC class I and II antigens in cultured human embryonic myoblasts following stimulation with gamma-interferon. *Immunol. Cell Biol.* **68**, 235–241.

Barr E. & Leiden J.M. 1991 Systemic delivery of recombinant protcins by genetically modified myoblasts. *Science, Wash.* **254**, 1507–1509.

Baumbach, L.L., Chamberlain, J.S., Ward, P.A., Farwell, N.J. & Caskey, C.T. 1989 Molecular and clinical correlations of deletions leading to Duchenne and Becker muscular dystrophies. *Neurology* **39**, 465–474.

Beggs, A.H., Koenig, M., Boyce, F.M. & Kunkel L.M. 1990 Detection of 98% of DMD/BMD gene deletions by polymerase chain reaction. *Hum. Genet.* **86**, 45–48.

Bodine, D.M., McDonagh, K.T., Seidel, N.E. & Nienhuis, A.W. 1991 Survival and retrovirus infection of murine hematopoietic stem cells *in vitro*: effects of 5-FU and method of infection. *Expl Hematol.* **19**, 206–212.

Brooke, M.H., Fenichel, G.M., Griggs, R.C. *et al.* 1989 Duchenne muscular dystrophy: patterns of clinical progression and effects of supportive therapy. *Neurology* **39**, 475–481.

Bulman, D.E., Gangopadhyay, S.B., Bebchuck, K.G., Worton, R.G. & Ray P.N. 1991 Point mutation in the human dystrophin gene: identification through Western blot analysis. *Genomics* **10**, 457–460.

Chamberlain, J.S., Gibbs, R.A., Ranier, J.E., Nguyen, P.N. & Caskey, C.T. 1988 Deletion screening of the Duchenne muscular dystrophy locus via multiplex DNA amplification. *Nucl. Acids Res.* **16**, 11141–11156.

Chamberlain, J.S., Gibbs, R.A., Ranier, J.E. & Caskey, C.T. 1989 Multiplex PCR for the diagnosis of Duchenne muscular dystrophy. In *PCR protocols: a guide to methods and applications* (ed. M. A. Innis, D. H. Getland, J. J. Sninsky & T. J. White), pp. 272–281. San Diego, California: Academic Press.

Clemens, P.R. & Caskey, C.T. 1992 Duchenne muscular dystrophy. In *Current neurology* (ed. S. H. Appel), pp. 1–22. Chicago, Illinois: Mosby-Year Book.

Clemens, P.R., Fenwick, R.G., Chamberlain, J.S. *et al.* 1991 Carrier detection and prenatal diagnosis in Duchenne and Becker muscular dystrophy families using dinucleotide repeat polymorphisms. *Am. J. hum. Genet.* **49**, 951–960.

Clemens, P.R., Ward, P.A., Caskey, C.T. & Fenwick, R.G. 1992 A premature chain termination mutation causing Duchenne muscular dystrophy. *Neurology*. (In the press.)

Cournoyer, D., Scarpa, M., Mitani, K. *et al.* 1991 Gene transfer of adenosine deaminase into primitive human hematopoietic progenitor cells. *Hum. Gene Ther.* **2**, 203–213.

Culver, K.W., Anderson, W.F. & Blease, R.M. 1991 Lymphocyte gene therapy. *Hum. Gene Ther.* **2**, 107–109.

Daddona, P.E., Shewach, D.S., Kelly, W.N., Argos, P., Markham, A.F. & Orkin, S.H. 1984 Human adenosine deaminase: cDNA and complete primary amino acid sequence. *J. biol. Chem.* **259**, 12101–12106.

Darras, B.T., Koenig, M., Kunkel, L.M. & Francke U. 1988 Direct method for prenatal diagnosis and carrier detection in Duchenne/Becker muscular dystrophy using the entire dystrophin cDNA. *Am. J. med. Genet.* **29**, 713–726.

Dhawan, J., Pan, L.C., Pavlath, G.K., Travis, M.A., Lanctot, A.M. & Blau, H.M. 1991 Systemic delivery of human growth hormone by injection of genetically engineered myoblasts. *Science, Wash.* **254**, 1509–1512.

Dick, J.E., Kamel-Reid, S., Murdoch, B. & Doedens, M. 1991 Gene transfer into normal human hematopoietic cells using *in vitro* and *in vivo* assays. *Blood* **78**, 624–634.

Dickson, G., Love, D.R., Davies, K.E., Wells, K.E., Piper, T.A. & Walsh F.S. 1991 Human dystrophin gene transfer: production and expression of a functional recombinant DNA-based gene. *Hum. Genet.* **88**, 53–58.

Eglitis, M.A. & Anderson, W.F. 1988 Retroviral vectors for introduction of genes into mammalian cells. *BioTechniques* **6**, 608–614.

England, S.B., Nicholson, L.V., Johnson, M.A., Forrest, S.M. & Love, D.R. 1990 Very mild muscular dystrophy associated with the deletion of 46% of dystrophin. *Nature, Lond.* **343**, 180–182.

Fenichel, G.M., Florence, J.M., Pestronk, A. et al. 1991 Long-term benefit from prednisone therapy in Duchenne muscular dystrophy. *Neurology* **41**, 1874–1877.

Ferrari, G., Rossini, S., Giavazzi, R. *et al.* An in vivo model of somatic cell gene therapy for human severe combined immunodeficiency. *Science, Wash.* **251**, 1363–1366.

Gilgenkrantz, H., Chelly, J., Lambert, M. *et al.* 1989 Analysis of molecular deletions with cDNA probes in patients with Duchenne and Becker muscular dystrophies. *Genomics* **5**, 574–580.

Gospe, S.M. Jr, Lazaro, R.P., Lava, N.S., Grootscholten, P.M., Scott, M.O. & Fischbeck, K.H. 1989 Familial X-linked myalgia and cramps: A nonprogressive myopathy associated with a deletion in the dystrophin gene. *Neurology* **39**, 1277–1280.

Griggs, R.C., Moxley, R.T. III, Mendell, J.R. *et al.* 1991 Prednisone in Duchenne dystrophy: a randomized, controlled trial defining the time course and dose response. *Arch. Neurol.* **48**, 383–388.

Grounds, M.D. 1991 Towards understanding skeletal muscle regeneration. *Path. Res. Pract.* **187**, 1–22.

Hohlfeld, R. & Engel, A.G. 1990*a* Human myoblasts as targets for allospecific cytotoxic T-lymphocytes and natural killer cells. *J. nurol. Sci.* **98** (Suppl.), 131.

Hohlfeld, R. & Engel, A.G. 1990*b* Induction of HLA-DR expression on human myoblasts with interferon-gamma. *Am. J. Pathol.* **136**, 503–508.

Hu, X.Y., Burghes, A.H., Bulman, D.E., Ray, P.N. & Worton R.G. 1989 Evidence for mutation by unequal sister chromatid exchange in the Duchenne muscular dystrophy gene. *Am. J. hum. Genet.* **44**, 855–863.

Hu, X.Y., Ray, P.N., Murphy, E.G., Thompson, M.W. & Worton R.G. 1990 Duplicational mutation at the Duchenne muscular dystrophy locus: its frequency, distribution, origin, and phenotype-genotype correlation. *Am. J. hum. Genet.* **46**, 682–695.

Huard, J., Bouchard, J.P., Roy, R. *et al.* 1991 Myoblast transplantation produced dystrophin-positive muscle fibres in a 16-year-old patient with Duchenne muscular dystrophy. *Clin. Sci.* **81**, 287–288.

Karpati, G., Pouliot, Y. & Carpenter S. 1988 Expression of immunoreactive major histocompatibility complex products in human skeletal muscles. *Ann. Neurol.* **23**, 64–72.

Karpati, G., Pouliot, Y., Carpenter, S. & Holland P. 1989*a* Implantation of nondystrophic allogenic myoblasts into dystrophic muscles of mdx mice produces 'mosaic' fibers of normal microscopic phenotype. In *Cellular and molecular biology of muscle development, UCLA Symposium on Molecular and Cellular Biology. New Series* (ed. L. Kedes & F. Stockdale), pp. 973–985. New York: Alan R. Liss, Inc.

Karpati, G., Pouliot, Y., Zubrzycka-Gaarn, E. *et al.* 1989*b* Dystrophin is expressed in mdx skeletal muscle fibers after normal myoblast implantation. *Am. J. Pathol.* **135**, 27–32.

Kilimann, M.W., Pizzuti, A., Grompe, M. & Caskey C.T. 1992 Point mutations and polymorphisms in the human dystrophin gene identified in genomic DNA sequences amplified by multiplex PCR. *Hum. Genet.* (In the press.)

Koenig, M., Hoffman, E.P., Bertelson, C.J., Monaco, A.P., Feener, C. & Kunkel, L.M. 1987 Complete cloning of the Duchenne muscular dystrophy (DMD) cDNA and preliminary genomic organization of the DMD gene in normal and affected individuals. *Cell* **50**, 509–517.

Koenig, M., Monaco, A.P. & Kunkel L.M. 1988 The complete sequence of dystrophin predicts a rod-shaped cytoskeletal protein. *Cell* **53**, 219–228.

Koenig, M., Beggs, A.H., Moyer, M. *et al.* 1989 The molecular basis for Duchenne versus Becker muscular dystrophy: correlation of severity with type of deletion. *Am. J. hum. Genet.* **45**, 498–506.

Kredich, N.M. & Hershfield, M.S. 1989 Immunodeficiency diseases caused by adenosine deaminase deficiency and purine nucleoside phosphorylase deficiency. In *The metabolic basis of inherited disease* (ed. C. R. Scriver, A. L. Beaudet, W. S. Sly & D. M. Valle), pp. 1045–1075. New York: McGraw-Hill.

Law, P.K., Goodwin, T.G. & Wang, M.G. 1988 Normal myoblast injections provide genetic treatment for murine dystrophy. *Muscle Nerve* **11**, 525–533.

Law, P.K., Bertorini, T.E., Goodwin, T.G. *et al.* 1990 Dystrophin production induced by myoblast transfer therapy in Duchenne muscular dystrophy. *Lancet* **336**, 114–115.

Law, P.K., Goodwin, T.G., Fang, Q.W. *et al.* 1991 Myoblast transfer therapy for Duchenne muscular dystrophy. *Acta Pediatr. Jpn.* **33**, 206–215.

Lee, C.C., Pearlman, J.A., Chamberlain, J.S. & Caskey C.T. 1991*a* Expression of recombinant dystrophin and its localization to the cell membrane. *Nature, Lond.* **349**, 334–336.

Lee, C.C., Pons, F., Jones, P.G. *et al.* 1991*b* Correction of dystrophic muscle by recombinant dystrophin in the *mdx* transgenic mouse. (Submitted.)

Levine, F. & Friedmann, T. 1991 Gene therapy techniques. *Curr. Opin. Biothch.* **2**, 840–844.

Love, D.R., Hill, D.F., Dickson, G. *et al.* 1989 An autosomal transcript in skeletal muscle with homology to dystrophin. *Nature, Lond.* **339**, 55–58.

Lubin, I., Faktorowich, Y., Lapidot, T. *et al.* 1991 Engraftment and development of human T and B Cells in mice after bone marrow transplantation. *Science, Wash.* **252**, 427–431.

Mantegazza, R., Hughes, S.M., Mitchell, D., Travis, M., Blau, H.M. & Steinman L. 1991 Modulation of MHC class II antigen expression in human myoblasts after treatment with IFN-gamma. *Neurology* **41**, 1128–1132.

Mendell, J.R., Moxley, R.T., Griggs, R.C. *et al.* 1989 Randomized, double-blind six-month trial of prednisone in Duchenne's muscular dystrophy. *New Engl. J. Med.* **320**, 1592–1597.

Miller, A.D. 1990 Progress toward human gene therapy. *Blood* **76**, 271–278.

Miller, R.G., Pavlath, G., Sharma, K. *et al.* 1992 Myoblast implantation in Duchenne muscular dystrophy: The San Francisco study. *Neurology* **42** (Suppl. 3), 189–190.

Morgan, J.E., Hoffman, E.P. & Partridge, T.A. 1990 Normal myogenic cells from newborn mice restore normal histology to degenerating muscles of the mdx mouse. *J. Cell Biol.* **111**, 2437–2449.

Mulligan, R.C. 1991 Gene transfer and gene therapy: principle, prospects, and perspective. In *Nobel Symposium 80: Etiology of Human Disease at the DNA Level* (ed. J. Lindsten & U. Pettersson), pp. 143–189. New York: Raven Press.

Multicenter Study Group 1992 Diagnosis of Duchenne and Becker muscular dystrophies by polymerase chain reaction. *J. Am. med. Ass.* **267**, 2609–2615.

Partridge, T.A., Morgan, J.E., Coulton, G.R., Hoffman, E.P. & Kunkel L.M. 1989 Conversion of mdx myofibres from dystrophin-negative to -positive by injection of normal myoblasts. *Nature, Lond.* **337**, 176–179.

Quantin, B., Perricaudet, L.D., Tajbakhsh, S. & Mandel, J.-L. 1992 Adenovirus as an expression vector in muscle cells *in vivo*. *Proc. natn. Acad. Sci. U.S.A.* **89**, 2581–2584.

Roberts, R.G., Bobrow, M. & Bentley, D.R. 1992 Point mutations in the dystrophin gene. *Proc. natn. Acad. Sci. U.S.A.* **89**, 2331–2335.

Salminen, A., Elson, H.F., Mickley, L.A., Fojo, A.T. & Gottesman M.M. 1991 Implantation of recombinant rat myocytes into adult skeletal muscle: a potential gene therapy. *Hum. Gene Ther.* **2**, 15–26.

Scarpa, M., Cournoyer, D., Muzny, D.M., Moore, K.A., Belmont, J.W. & Caskey, C.T. 1991 Characterization of recombinant helper retroviruses from Moloney-based vectors in ecotropic and amphotropic cell lines. *Virology* **180**, 849–852.

Sutherland, H.J., Eaves, C.J., Eaves, A.C., Dragowska, W. & Lansdorp, P.M. 1989 Characterization and partial purification of human marrow cells capable of initiating long term hematopoiesis *in vitro*. *Blood* **74**, 1563–1570.

Sutherland, H.J., Lansdorp, P.M., Henkelman, D.H., Eaves, A.C. & Eaves, C.J. 1990 Functional characterization of individual human hematopoietic stem cells cultured at limiting dilution on supportive marrow stromal layers. *Proc. natn. Acad. Sci. U.S.A.* **87**, 3584–3588.

Verma, I.M. & Naviaux, R.K. 1991 Human gene therapy. *Curr. Opin. Genet. Dev.* **1**, 54–59.

Ward, P.A., Hejtmancik, J.F., Witkowski, J.A. *et al.* 1989 Prenatal diagnosis of Duchenne muscular dystrophy: prospective linkage analysis and retrospective dystrophin cDNA analysis. *Am. J. hum. Genet.* **44**, 270–281.

Wells, D.J., Wells, K.E., Walsh, F.S. *et al.* 1992 Human dystrophin expression corrects the myopathic phenotype in transgenic *mdx* mice. *Hum. molec. Genet.* **1**, 35–40.

Yang, N.-S., Burkholder, J., Roberts, B., Martinell, B. & McCabe D. 1990 *In vivo* and *in vitro* gene transfer to mammalian somatic cells by particle bombardment. *Proc. natn. Acad. Sci. U.S.A.* **87**, 9568–9572.

11

Enhancing the efficiency of transgene expression

A. J. Clark, A. L. Archibald, M. McClenaghan, J. P. Simons, R. Wallace and C. B. A. Whitelaw

SUMMARY

Two strategies for enhancing gene expression in transgenic animals are described with particular reference to targeting expression to the mammary gland. Gene constructs in which the protein-encoding DNA sequences are contained within a genomic segment (comprising most or all of the natural introns of the corresponding gene) are shown to be expressed more efficiently than their intronless counterparts. Secondly, co-integrating an otherwise poorly expressed transgene in the vicinity of an actively expressed transgene can dramatically improve its efficiency of expression.

1. INTRODUCTION

The ability to manipulate the germline of animals offers a number of biotechnological opportunities. Recent interest has focused on the use of transgenic animals for the production of recombinant proteins, in particular human biomedical proteins. The approach generally taken has been to target expression to the mammary gland and produce the desired protein in milk. Livestock such as sheep or cows, or even pigs, synthesize and secrete large amounts of protein in their milk during lactation. Using gene transfer techniques this synthetic capacity can be harnessed to produce other valuable proteins, thus offering a relatively cheap method of production compared to conventional fermentation-scale mammalian cell culture (Clark *et al.* 1987; Whitelaw & Clark 1989).

Milk protein genes from a variety of species have been isolated and shown to function appropriately in the mammary gland of transgenic mice (Simons *et al.* 1987; Vilotte *et al.* 1989; Bayna & Rosen 1990). Regulatory sequences from these and other milk protein genes have been used to target expression of human proteins in transgenic mice (Gordon *et al.* 1987; Yu *et al.* 1989; Archibald *et al.* 1990; Meade *et al.* 1990; DiTullio *et al.* 1992), rabbits (Buhler *et al.* 1990) sheep (Clark *et al.* 1989; Wright *et al.* 1991) and goats (Ebert *et al.* 1991). Proteins produced by this route are biologically active demonstrating that the mammary gland is capable of carrying out the appropriate post-translational modifications. In many cases, however, the efficiency of expression of these hybrid transgenes was shown to be low, both in terms of the frequency and levels of expression. The inefficient and unpredictable mode of expression of transgenes is a major constraint to the adoption of this technology. This is particularly the case when transgenic livestock are required for large-scale production, given that the efficiency of gene transfer in these species is low and the procedures involved both expensive and time-consuming (Pursel *et al.* 1989; Clark *et al.* 1991). It is clearly important to understand more fully the processes that control the expression of transgenes and, if possible, manipulate them to improve their efficiency.

To target expression to the mammary gland we have used sequences derived from the sheep milk protein gene β-lactoglobulin (BLG, Ali & Clark 1988). The unmodified gene, itself, is expressed efficiently and specifically in transgenic mice (Simons *et al.* 1987; Harris *et al.* 1991). We have shown that essential regulatory elements are located within the proximal 406 b.p. of 5′ flanking sequences (Whitelaw *et al.* 1992) and that this region contains the binding sites for a putative mammary gland-specific transcription factor (Watson *et al.* 1991). BLG transgenes comprising this region, the transcription unit and 3′ flanking sequences are expressed in a position-independent manner: they are invariably expressed, and the levels of expression are related to the copy number of integrated transgenes. In this paper we describe two approaches for improving the expression of hybrid genes containing regulatory element from the BLG gene. Firstly, we demonstrate a requirement for introns for efficient expression and, secondly, we show that chromosomal position effects at the site of integration can be manipulated to enhance expression.

2. INTRON RESCUE

The DNA sequences encoding a particular protein of interest are often only available as cDNA sequences. This may reflect the original cloning experiments or the fact that many candidate proteins are encoded by very large genes (for example, the gene encoding human factor VIII is about 180 kb in length; Gitschier *et al.* 1984) and not conveniently manipulated by conventional recombinant DNA techniques. As such, cDNA segments lack introns and although no specific

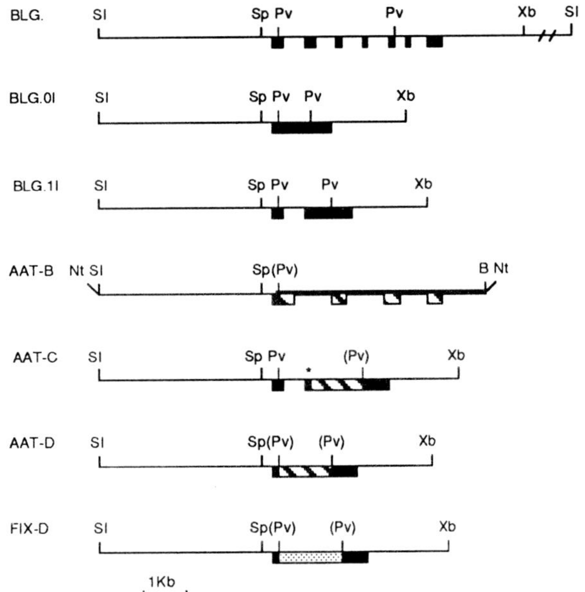

Figure 1. Structure of BLG and BLG hybrid genes introduced into transgenic mice. Thin line, BLG flanking sequences and introns; thick line α_1AT flanking sequences and introns; filled boxes, BLG exons and/or cDNA; hatched boxes, α_1AT exons and/or cDNA; lightly shaded box, fIX cDNA; * inserted termination codon. Relevant restriction sites; B, *Bam*HI; Nt, *Not*I; Sl, *Sal*I; Sp, *Sph*I; Pv, *Pvu*II; Xb, *Xba*I. BLG sites lost during ligation are shown in brackets. Data from Whitelaw *et al.* (1991).

role in controlling gene expression has been ascribed to them, their absence is a common feature of many the poorly performing transgenes that we and other workers have described.

Figure 1 shows a number of different BLG-derived constructs designed to evaluate whether introns are required for efficient expression. BLG, BLG-0I and BLG-1I are derived solely from the BLG gene and comprise the unmodified gene, an intronless derivative and an intron-1 containing derivative, respectively. AAT-B, AAT-C and AAT-D are hybrid genes comprising various configurations of human alpha-1-antitrypsin (α_1AT) encoding sequences with BLG sequences. AAT-B comprises ~4.0 kb of 5′ flanking sequences of the BLG gene fused to a genomic minigene encoding human α_1AT. In this construct three of the four 'natural' introns present in the human α_1AT gene are retained. AAT-D comprises an intronless BLG/α_1AT hybrid gene in which the 5′ and 3′ segments of BLG have been linked to a cDNA segment encoding human α_1AT. AAT-C comprises the same α_1AT cDNA segment but, by analogy with BLG-1I, the first intron of BLG has been included upstream. Finally, FIX-D comprises a cDNA segment encoding human factor IX (fIX) fused to the 5′ and 3′ flanking sequences of BLG; it contains no introns and is analogous to BLG-0I and AAT-D.

These constructs were introduced into transgenic mice by microinjection into the pronuclei of fertilized eggs from superovulated (C57BL/6 × CBA) F1 female mice (Simons *et al.* 1987). Expression was analysed in G_0 or G_1 (when the G_0 was male) female transgenic mice. Mid-lactation females were milked and milk samples were analysed by SDS-PAGE, Western blotting and radioimmunoassay as described previously (Simons *et al.* 1987; Archibald *et al.* 1990). After milking, the mice were killed and total RNA was prepared from the mammary gland as well as a small

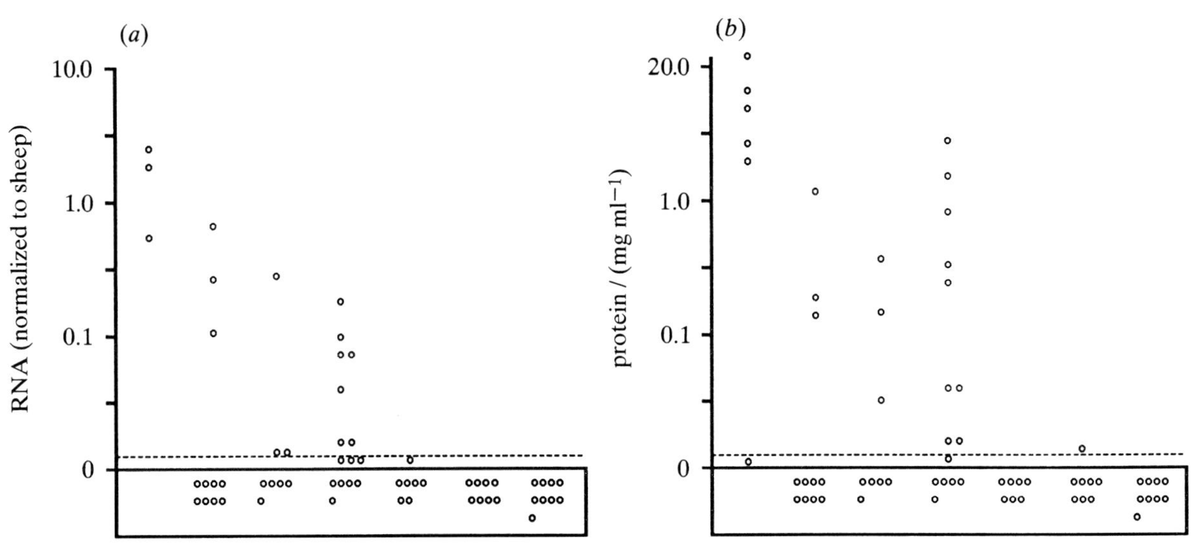

Figure 2. Comparison of transgene expression. Steady state BLG and α_1AT mRNA levels in the mammary gland (*a*) and protein levels in mouse milk (*b*) are compared for the various constructs shown in figure 1. Each circle corresponds to the levels determined for a G_0 or G_1 female transgenic mouse; values below the dotted line were detectable but could not be accurately determined. Note that RNA levels were not determined for all BLG mice. Statistical comparison of expression frequencies (Fisher's Exact Test) determined that BLG-0I and BLG-1I were significantly different from BLG ($p < 0.01$ and $p < 0.05$, respectively); AAT-D and FIX-D were significantly different from AAT-B ($p < 0.05$ and $p < 0.01$, respectively) while AATC was nearly significantly different from AAT-B ($p < 0.06$); no significant difference was detected between BLG and AAT-B. For full details see Whitelaw *et al.* (1991).

selection of other tissues. RNA levels in the mammary gland were quantified by probing dot blots loaded with serial dilutions using sheep mammary and human liver RNAs as controls.

Figure 2 summarizes the data obtained from these experiments. A clear difference emerges when constructs comprising most or all of their native introns (BLG and AAT-B) are compared with constructs that lack introns (BLG-0I, AAT-D, FIX-D) or contain just one BLG intron (BLG-1I, AAT-C). Thus, all mice (or lines) that carried the unmodified BLG gene expressed it, and in five of the six cases high levels of BLG (greater than 1 mg ml^{-1}) were obtained in the milk. The efficient expression of native BLG transgenes has been confirmed in an additional set of experiments involving a further 24 transgenic mice (or lines) carrying comprising various amounts of 5′ flanking DNA, but otherwise identical to the construct shown in figure 1. In every case significant expression was observed and, furthermore, the level of expression showed a significant correlation with the copy number of integrated transgenes (Whitelaw *et al.* 1992). By contrast only three of the eight mice (or lines) carrying the intronless gene, BLG-0I, expressed the transgene. In those mice that did express the average level of expression was approximately one twentieth the level obtained with the unmodified BLG transgene. A similar result was obtained with BLG 1I, which contains only the first BLG intron. Only three of the 11 mice (or lines) expressed this construct and the average level of expression in these animals was, again, lower than was observed with BLG.

A similar difference in the efficiency of expression was observed when the genomic α_1AT construct AAT-B was compared to the the two α_1AT cDNA constructs AAT-D and AAT-C: ten of the 15 AAT-B mice (or lines) expressed this construct in the mammary gland. The highest expressing line produced milk concentrations of human α_1AT of between 7 and 8 mg ml^{-1} and, generally, there was a good correlation between the steady-state mRNA levels and the level of protein detected in milk (Whitelaw *et al.* 1991). The level of expression of this transgene, however, was more variable than observed with BLG and no relationship was observed between expression levels and copy number. In contrast to AAT-B, only one of the eight AAT-D transgenics expressed detectable α_1AT protein and only one of the eight AAT-C animals expressed detectable RNA (figure 2). No expression was observed in the nine mice (or lines) which carried the fIX cDNA construct FIX-D.

Transcript sizes and tissue specificity were determined in expressing animals by Northern blotting experiments. BLG and BLG-minigenes gave rise to the appropriate sized transcripts and expression was restricted to the mammary gland, showing that BLG introns do not contain sequences required to limit tissue specificity. Northern blotting of RNA samples from AAT-B mice also showed transcripts of the expected size in the mammary gland of the ten expressing mice (or lines). By contrast to BLG constructs, four of these mice (or lines) were also shown to express AAT-B transcripts in the salivary gland. Additionally, four of the five animals that did not express AAT-B in the mammary gland showed detectable expression in the salivary gland. We do not understand at present why the AAT-B transgene is expressed at such a high frequency in the salivary gland. It is possible that downstream BLG sequences contain a salivary gland-specific negative regulatory element or that regulatory elements within the α_1AT genomic sequences direct expression to this tissue.

Other workers have also investigated the requirement for intronic sequences in transgenic mice (Brinster *et al.* 1988; Palmiter *et al.* 1991; Choi *et al.* 1991). In these studies intron-containing and intronless gene constructs were targeted to various tissues in transgenic mice. Introns were shown to improve expression in a variety of tissues showing that this may be a general requirement for the efficient expression of transgenes.

We do not fully understand the mechanism(s) whereby introns enhance transgene expression. The failure of cDNA-based hybrid genes to be expressed at high frequency in the mammary gland does not appear to due to a sinple requirement for splicing because AAT-C (which contains the first intron of BLG) was expressed as poorly as its intronless counterpart. Secondly, BLG-1I (which contains the first BLG intron) was expressed no more efficiently than its intronless counterpart, BLG 0I.

The performance of BLG hybrid genes and BLG minigenes in transgenic mice is reminiscent of the behaviour of transgenes carrying weak promoters (see, for example, Bonnerot *et al.* (1990); Allen *et al.* (1988)) which are particularly sensitive to chromosomal position effects. All the constructs described in this study comprise identical 5′ flanking regions and we have postulated that it is the interaction between the 5′ promoter sequences and *cis* elements downstream that determines how effectively chromosomal position effects are overcome and, hence, how efficiently a given construct will function (Whitelaw *et al.* 1991). Intron containing constructs such as BLG and AAT-B may function relatively efficiently because they contain such *cis* acting elements rather than any obligatory requirement for splicing. One possibility is that these putative *cis*-acting sequences enable a transgene to adopt the appropriate chromatin configuration and allow access to the soluble transcription factors which then drive expression from the 5′ promoter segment. The precise nature of these sequences is not known: they may be discretly organized or distributed throughout the transcription unit.

From a practical point of view a general requirement for introns has a number of implications for transgene design. Limitations are imposed on the construction of transgenes particularly those in which the structural region encompasses large regions of chromosomal DNA (e.g. factor VIII; Gitschier *et al.* 1984). Genomic minigenes such as AAT-B do function quite efficiently and, therefore, it should be possible to produce efficiently expressing constructs that comprise a mixture of cDNA and genomic sequences. Nevertheless, given our lack of understanding as to how introns enhance gene expression the choice of which to include in any given gene construct will not be

Table 1. *Production of β-lactoglobulin-alpha$_1$-antitrypsin (BAD) and β-lactoglobulin-human factor IX (BIX) transgenic mice*

(Tail DNA samples from G_0 mice were screened initially by the polymerable chain reaction (PCR) using primers which detect all three transgenes. PCR-positive samples were further analysed by Southern blotting analysis, using BLG-specific, AATD-specific and FIXD-specific probes (BLG, β-lactoglobulin gene; AATD, α-1-antitrypsin-D hybrid gene; FIXD, human factor IX-D hybrid gene). Data from Clark *et al.* (1992).)

		double transgenics		single transgenics		
	total G_0 mice	BLG + AATD	BLG + FIXD	BLG	AATD	FIXD
BAD	173	11		1	8	
BIX	135		20	10		0

straightforward and may require the *ad hoc* evaluation of a number of different alternatives.

3. TRANSGENE RESCUE

Poorly expressed transgenes (such as those lacking the appropriate introns) appear to be highly influenced by chromosomal position effects which are presumed to act at the site of integration. We reasoned, therefore, that engineering the site of integration rather than the construct *per se* might provide a strategy to rescue poorly expressed transgenes. For targeting expression to the mammary gland the vicinity of an actively expressed milk protein could constitute a position which would rescue the expression of, otherwise, poorly expressed constructs. Gene targeting by homologous recombination in embryonic stem cells (Capecchi 1989) would, in principle, enable the insertion of constructs adjacent to endogenous milk protein genes, for example, within the cluster of casein genes. Gene targeting into the germline is, however, technically difficult and, for the present, limited to mice since no proven ES cells have yet been isolated from other mammalian species. We have favoured an

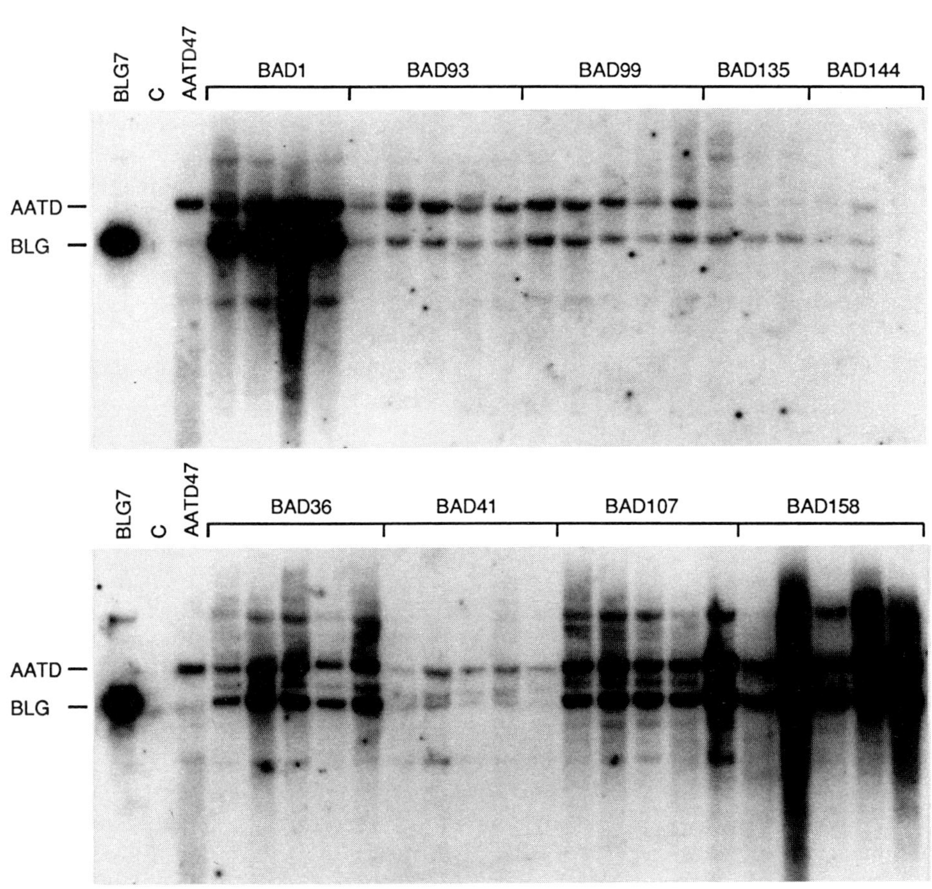

Figure 3. Co-segregation of BLG and AAT-D in BAD lines. Tail DNA samples from a number of individuals from each BAD line (BAD 1–BAD 158) and controls were cleaved with *Eco*RI and analysed by electrophoresis and Southern blotting. The filters were probed with 5′BLG sequences which hybridize to both AAT-D and BLG transgenes. BLG7, DNA from a single BLG transgenic mouse; C, control non-transgenic DNA; AATD47, DNA from a single AAT-D transgenic line. The AAT-D-specific and the BLG-specific fragments are indicated. Data from Clark *et al.* (1992).

alternative approach that involves the co-injection of different DNA constructs into pro-nuclei, since this can result in the co-integration of the DNAs at a single site (Storb *et al.* 1986; Behringer *et al.* 1989). The BLG gene is efficiently expressed in transgenic mice and, moreover, seems to contain all the sequences necessary for overcoming chromosomal position effects (Whitelaw *et al.* 1992). Therefore, to test the rescue hypothesis, transgenic mice were produced by co-injection of the BLG gene with either the AAT-D or FIX-D constructs which are expressed very inefficiently when integrated alone (figure 2).

In the first experiment the BLG and AAT-D constructs (figure 1) were co-injected into mouse eggs in a 1:1 molar ratio. The constructs were not ligated prior to microinjection. The co-integration frequency was high and more than 50% of the G_0 transgenic mice obtained had integrated both BLG and AAT-D transgenes (table 1). From the 11 G_0 BLG/AAT-D (BAD) mice, nine transgenic lines were established. Southern blotting experiments carried out on a number of G_1 progeny showed the same pattern of AAT-D and BLG-specific fragments within each line (figure 3). Thus in each line the BLG and AAT-D transgenes have co-segregated, suggesting a single site of integration. One exception was line 93 in which an additional fragment was observed in some of the G_1 mice, most probably resulting from a second site of integration. More detailed Southern blotting analyses of DNA prepared from the BAD mice identified junction fragments between BLG and AAT-D confirming co-integration of the two transgenes. Transgene copy number varied widely between lines although within a line, BLG and AAT-D copy numbers were similar, reflecting the molar ratio of the DNAs injected (figure 3).

Milk was collected from transgenic mid-lactation G_0 or G_1 females and assayed by ELISA. Human α_1AT was detected in the milk of mice from seven of the nine lines. The concentrations of the foreign protein ranged from ~ 1 μg ml^{-1} to over 600 μg ml^{-1} (figure 4). Total RNA was prepared from the mammary gland and a small selection of other tissues and probed for AAT-D transcripts (figure 5). The expected transcripts were detected in mammary gland RNA in six of the nine BAD lines and their steady-state levels correlated approximately with the concentration of human α_1AT in milk (we were unable to detect AAT-D transcripts in the lowest expressing BAD line). No expression of AAT-D was detected in other tissues. The efficiency of expression of AAT-D when cointegrated with BLG contrasts markedly with expression of the singly integrated AAT-D construct. In the latter case low level expression was detected in only one of eight mice (or lines), whereas in the BAD mice AAT-D expression was detected in seven of the nine lines, two of which expressed more than 100 μg ml^{-1} of the foreign protein.

The second rescue experiment was carried out using the FIX-D construct (figure 1). In this experiment the BLG fragment was injected in a threefold molar excess over the FIX-D fragment. Again the efficiency of co-integration was high and 20 of the 30 G_0 animals obtained carried both FIX-D and BLG transgenes (table 1). From 13 of the BLG/FIX-D (BIX) transgenic mice germline transmission was obtained from 11. As for the BAD mice the two genes co-segregated and the majority were co-integrated at one site only. By contrast to the BAD lines, the BLG copy numbers in the BIX lines were higher than those of FIX-D, indicating that the composition of the integration sites reflects the ratio of the input DNAs microinjected (not shown).

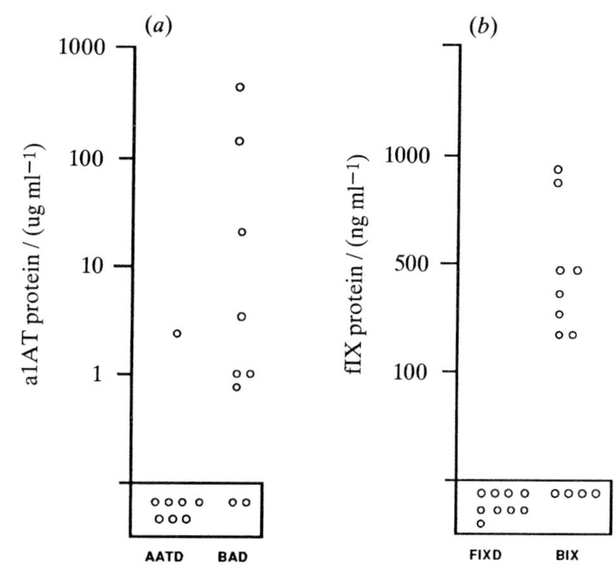

Figure 4. Expression of co-integrated transgenes. α_1AT (*a*) and fIX (*b*) protein levels (measured by ELISA) in BAD and BIX mice were compared to the corresponding single transgenics carrying AAT-D and FIX-D. The circles correspond to the levels determined for independently-derived G_0 or G_1 females. For full details see Clark *et al.* (1992).

Milk and RNA samples were collected from mid-lactation females. Factor IX was detected by ELISA in milk from eight of 12 BIX transgenic mice, although the concentrations were low ranging from 0.2–1 μg ml^{-1} (figure 4). FIX-D transcripts were detected specifically in the mammary gland (figure 5) in samples from ten of the 12 mice (or lines) analysed. Steady-state levels varied considerably between mice (or lines) but those in which fIX was detected in the milk (eight of 12) exhibited relatively abundant transcripts; these were considerably higher (up to 60-fold) than the level of human fIX mRNA in human liver, the normal site of expression of factor IX. The FIX-D transcripts were, however, approximately 400 nt shorter than predicted. Additional Northern blotting experiments (not shown) demonstrated that this was due to an internal deletion of the transcripts, presumably due to aberrant splicing. Thus the low concentration of fIX detected in milk may result from the translation of a small proportion of non-deleted FIX-D transcripts, the innefficient translation of the truncated mRNA or the innefficient secretion of an aberrant fIX. Notwithstanding, the problems of mRNA processing, co-integration of FIX-D with BLG dramatically enhanced its efficiency of expression: in none of nine single transgenics was expression detected

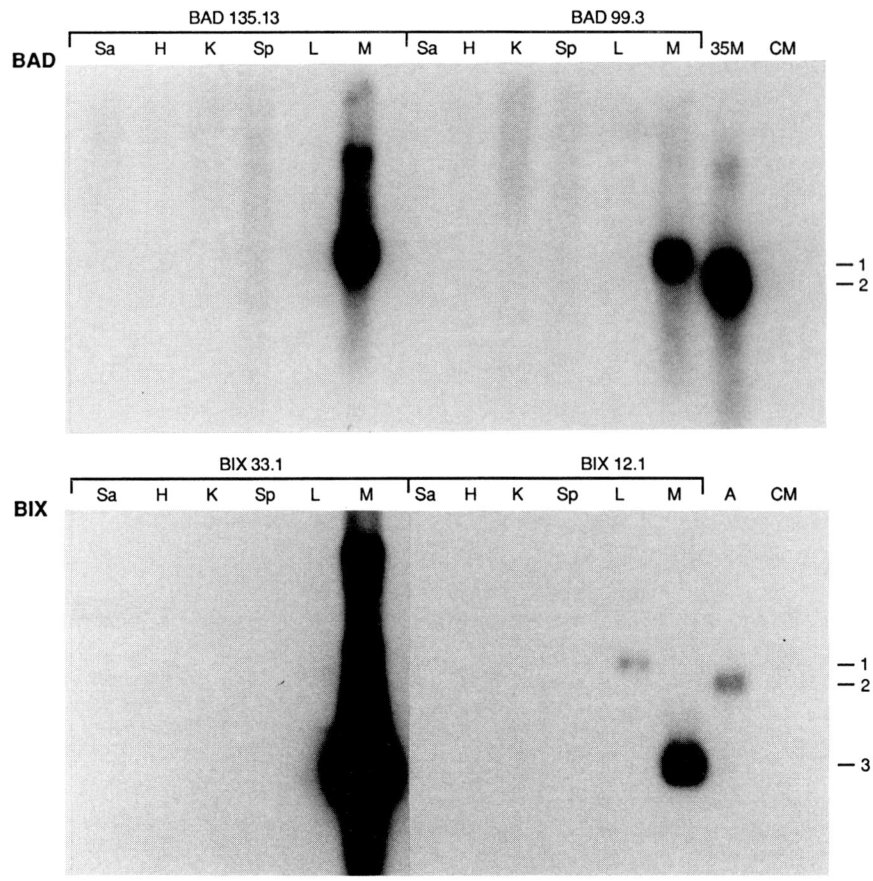

Figure 5. Tissue-specific expression in BAD and BIX mice. Total RNA samples (10 μg) from controls, BAD, and BIX mice were analysed by Northern blotting. Top panel, BAD RNA samples: CM, control mouse mammary RNA; 35M, mammary gland RNA from mouse AAT-B35 (Archibald *et al.* 1990); M, mammary gland; L, liver ; Sp, spleen; K, kidney; H, heart; Sa, salivary gland. The ~1600 nt AAT-D (1) and ~1400 nt AAT-B (2) transcripts are indicated at the right-hand side of the panel. Bottom panel, BIX RNA samples: A, mammary gland RNA from FIXA51 a transgenic mouse expressing FIXA, a fIX construct we have previously shown to be expressed in transgenic sheep (Clark *et al.* 1989); other tracks as for the top panel. The ~2600 nt endogenous mouse fIX transcript (1), ~2400 nt FIXA transcript (2) and ~1450 nt FIX-D (3) transcripts are indicated at the right-hand side of the panel. For further details see Clark *et al.* (1992).

(figure 1), whereas the majority (ten of 12) BIX mice (or lines) expressed the construct.

The data from both BAD and BIX experiments indicates that engineering the site of integration for a particular construct can be used as a strategy to enhance expression. At present we do not understand the mechanism whereby co-integration with BLG rescues the expression of an otherwise poorly expressed construct. Possible mechanisms include the abrogation of negative position effects, whereby sequences within the BLG gene insulate the hybrid genes from the chromosomal sequences flanking the integration site (Kellum & Schedl 1991). Alternatively, it is possible that the adjacent BLG gene provide positive effects on expression. Such positive effects could operate via interactions of the BLG gene and the rescued genes through enhancer or LCR-like activity (Grosveld *et al.* 1987). Alternatively, the open chromatin structure surrounding the expressed BLG gene may spread to encompass the flanking sequences, including the rescued genes.

4. DISCUSSION

We have described two studies on improving the efficiency of gene expression in transgenic animals. The first shows that the inclusion of introns in gene constructs can improve their performance. In our hands constructs containing most or all of their natural introns were expressed more efficiently than their intronless counterparts. Inclusion of one BLG intron did not improve the efficiency of expression indicating that a requirement for splicing was not the reason for the poor performance of these constructs. Other workers (Brinster *et al.* 1988; Palmiter *et al.* 1991; Choi *et al.* 1991) have demonstrated similar results with a variety of constructs and have also suggested that insertion of a heterologous intron between the promoter and coding region can enhance expression. From a practical point of view, all these studies support the use of genomic constructs wherever possible for transgenic experiments, with the next best strategy being the use of a minigene in which as much as possible of the genomic sequence is retained but large intronic regions are removed. This latter strategy is exemplified in this current work by construct AAT-B. Not only does this minigene construct function efficiently in transgenic mice it also performs very well in transgenic sheep. Each of the four G_0 females expressed human α_1AT in milk; all expressed at levels above 1 mg ml^{-1} with the highest expressor producing a sustained level of ~35 mg ml^{-1} (Wright *et al.* 1991).

If the appropriate genomic or genomic minigene constructs are not readily available then the second strategy presented here, transgene rescue, may provide a solution for enhancing transgene expression. Co-injection provides a simple method for introducing two gene constructs into the germline but suffers from the problem that complex and variable transgene arrays are generated at the different sites of integration. The variable structure of these arrays may be partly or wholly resposible for the wide variation of expression levels between different lines and suggests that there is an optimal configuration of the two transgenes. Building double-contructs prior to injection could provide more control over the structure of the arrays by ensuring that the optimal configuration is present. Finally, although the present data are limited to targeting gene expression to the mammary gland it seems likely that transgene rescue will work in other tissues, given that a suitable 'rescuing' gene is available.

It is tempting to speculate that intron rescue and transgene rescue involve a common mechanism(s) at the chromosomal level, for example, by the generation of the appropriate chromatin configurations. Whatever the mechanisms involved it is clear that the design of efficiently expressing gene constructs for gene transfer entails more than simply linking a 'strong' promoter to the appropriate coding sequences. As well as providing some practical solutions to transgene design the results presented in these two studies suggest that other, as yet poorly understood, processes are critical for efficient gene expression.

REFERENCES

Ali, S. & Clark, A.J. 1988 Characterisation of the gene encoding ovine β-lactoglobulin. *J. molec. Biol.* **199**, 415–426.

Allen, N.D., Cran, D.G., Barton, S.G., Hettle, S., Reik, W. & Surani, M.A. 1988 Trangenes as probes for active chromosomal domains in mouse development. *Nature, Lond.* **333**, 852–855.

Archibald, A.L., McClenaghan, M., Hornsey, V., Simons, J.P. & Clark, A.J. 1990 High level expression of biologically active human α_1-antitrypsin in transgenic mice. *Proc. natn. Acad. Sci. U.S.A.* **87**, 5178–5182.

Bayna, E.M. & Rosen, J.M. 1990 Tissue-specific, high-level expression of the rat whey acidic protein gene in transgenic mice. *Nucl. Acids Res.* **18**, 2877–2885.

Behringer, R.R., Ryan, T.M., Reilly, M.P., Asakura, T., Palmiter, R.D. & Brinster, R.L. 1989 Synthesis of functional hemoglobin in transgenic mice. *Science, Wash.* **245**, 971–973.

Bonnerot, C., Grimber, G., Briand, P. & Nicolas, J.-F. 1990 Patterns of expression of position dependent integrated transgenes in mouse embryos. *Proc. natn. Acad. Sci. U.S.A.* **87**, 331–6335.

Brinster, R.L., Allen, J.M., Behringer, R.R., Gelinas, R.E. & Palmiter, R.D. 1988 Introns increase transcriptional efficiency in transgenic mice. *Proc. natn. Acad. Sci. U.S.A.* **85**, 836–840.

Buhler, T. A., Bruyere, T., Went, D.F., Stranzinger, G. & Burki, K. 1990 Rabbit β-casein promoter directs secretion of human interleukin-2 into milk of transgenic rabbits. *BioTechnology* **8**, 140–143:

Capecchi, M.R. 1989 Altering the genome by homologous recombination. *Science, Wash.* **244**, 1288–1292.

Choi, T., Huang, M., Gorman, C., & Jaenisch, R. 1991 A generic intron increases gene expression in transgenic mice. *Molec. Cell Biol.* **11**, 3070–3074.

Clark, A.J., Simons, J.P., Wilmut, I. & Lathe, R. 1987 Pharmaceuticals from transgenic livestock. *Trends Biotech.* **5**, 20–24.

Clark, A.J., Bessos, H., Bishop, J.O. *et al.* 1989 Expression of human anti-hemophilic factor IX in the milk of transgenic sheep. *Bio Technology.* **7**, 487–492.

Clark, A.J., Archibald, A.L., McClenaghan, M. *et al.* 1991 The germline manipulation of livestock: progress during the past five years. *Proc. N.Z. Soc. Anim. Prod.* **50**, 167–180.

Clark, A.J., Cowper, A., Wallace, R. *et al.* 1992 Rescuing transgene expression by co-integration. *Bio Technology* **10**, 1450–1456.

DiTullio, P., Cheng, S.H., Marshall, J. *et al.* 1992 Production of cystic fibrosis transmembrane conductance regulator in the milk of transgenic mice. *BioTechnology* **10**, 74–77.

Gitschier, J., Wood, W.I., Goralka, T.M. *et al.* 1984 Characterisation of the human factor VIII gene. *Nature, Lond.* **312**, 326–330.

Gordon, K., Lee, E., Vitale, J.E. *et al.* 1987 Production of human tissue plasminogen activator in transgenic mouse milk. *Bio Technology* **5**, 1183–1187.

Grosveld, F. van Assendelft, G.B., Greaves, D.R. & Kollias G. 1987 Position independent, high level expression of the human beta-globin gene. *Cell* **51**, 975–985.

Harris, S., McClenaghan, M., Simons, J.P., Ali, S. & Clark, A.J. 1991 Developmental regulation of the sheep β-lactoglobulin gene in transgenic mice. *Devl. Genet.* **12**, 299–307.

Kellum, R. & Schedl, P. 1991 A position effect assay for boundaries of higher order chromosomal domains. *Cell* **64**, 941–950.

Meade, H., Gates, L., Lacy, E. & Lonberg, N. 1990 Bovine α_{s1} sequences direct high level expression of active human urokinase in mouse milk. *BioTechnology* **8**, 443–446.

Palmiter, R.D., Sandgren, E.P., Avarbock, M.R., Allen, J.M & Brinster, R.L. 1991 Heterologous introns can increase expression in transgenic mice. *Proc. natn. Acad. Sci. U.S.A.* **88**, 478–482.

Pursel, V.G., Pinkert, C.A., Miller, K.F., Bolt, D.J., Campbell, R.G., Palmiter, R.D., Brinster, R.L. & Hammer, R.E. 1989 Genetic engineering of livestock. *Science, Wash.* **244**, 1281–1289.

Storb, U., Pinkert, C., Arp, B. *et al.* 1986 Transgenic mice with mu and kappa genes encoding antiphosphorylcholine. *J. exp. Med.* **164**, 627–641.

Simons, J.P., McClenaghan, M. & Clark, A.J. 1987 Alteration of the quality of milk by expression of sheep β-lactoglobulin in transgenic mice. *Nature, Lond.* **328**, 530–532.

Vilotte, J.-L., Soulier, S., Stinnakre, M.-G., Massoud, M. & Mercier, J.-C. 1989 Efficient tissue-specific expression of bovine α-lactalbumin in transgenic mice. *Eur. J. Biochem.* **186**, 43–48.

Watson, C.J., Gordon, K.E., Robertson, M. & Clark, A. J. 1991 Interaction of DNA binding proteins with a milk protein gene promoter in vitro: identification of a mammary gland-specific factor. *Nucl. Acids Res.* **19**, 6603–6610.

Whitelaw, C.B.A. & Clark, A.J. 1989 Animal bioreactors. *AgBiotech News Inform.* **1**, 701–705.

Whitelaw, C.B.A., Archibald, A.L., Harris, S., McClenaghan, M., Simons, J.P. & Clark, A.J. 1991 Targeting expression to the mammary gland; intronic sequences can enhance the efficiency of gene expression in transgenic mice. *Transgenic Res.* **1**, 3–13.

Whitelaw, C.B.A., Harris, S., McClenaghan, M., Simons, J.P. & Clark, A.J. 1992 Position-independent expression of the ovine β-lactoglobulin gene in transgenic mice. *Biochem. J.* **286**, 31–39.

Wright, G., Carver, A., Cottom, D. *et al.* 1991 High level expression of active human alpha-1-antitrypsin in the milk of transgenic sheep. *BioTechnology.* **9**, 830–834.

Yu, S.-H., Deen, K.C., Lee, E. *et al.* 1989 Functional human CD4 protein produced in the milk of transgenic mice. *Molec. biol. Med.* **6**, 255–261.

12

Improved agronomic and quality traits in transgenic crops: recent advances

M. E. Fromm, D. M. Stark, G. D. Austin and F. J. Perlak

SUMMARY

The potential for genetic engineering of plants to produce new and useful traits in crops has now been confirmed. Although numerous examples exist in the literature of agronomically important traits, the examples of quality improvements are rare but increasing. This article describes several new agronomic and quality traits produced at the Monsanto Plant Science Division. Tomatoes have been produced that have reduced ethylene levels which result in delayed fruit ripening. Delayed ripening contributes to an extended shelf life of the tomatoes. Higher starch potatoes add value both by the increased yield of starch and by decreasing oil absorption during frying. Cotton that expresses the *Bacillus thuringiensis* gene controls lepidopteran insect feeding damage as effectively as chemical insecticides. These genetically improved plants are valuable additions to modern agriculture and represent clear demonstrations of the types of improvements that are likely in the future.

1. IMPROVING TOMATO QUALITY BY CONTROLLING FRUIT RIPENING

The average annual per capita consumption of fresh tomatoes is eighteen pounds and is increasing at approximately 2% per year. Retail and food service sales of fresh tomatoes have surpassed those of potatoes and lettuce and are valued at greater than $3.5 billion per year. However, the quality of the average store bought tomato still fails to meet consumer expectations and typically is described as having poor flavour and a mealy texture.

The inferior flavour of current fresh tomatoes is a consequence of the stage of fruit development at which time the fruit are harvested. Flavour development correlates to length of time on the vine (Kader *et al.* 1977). To provide tomato fruit year round, agricultural practices were modified to incorporate mature green harvested tomato fruit. Tomato fruit are harvested at the mature green stage and treated with ethylene to facilitate ripening off the vine. Mature green fruit possess the physical properties needed within a system designed to efficiently handle large volumes of tomatoes. Harvesting at a mature green stage also provides a fruit with sufficient shelf life (length of time which fruit are marketable) to enable processing and distribution of fruit over large geographic areas. Mature green fruit, however, do not obtain optimum flavour and when the fruit are allowed to develop full flavour by remaining on the vine longer, they lose the handling properties required for cost effective distribution of the tomato fruit. An additional problem with harvesting mature green fruit pertains to the difficulty in distinguishing a mature green fruit from an immature green fruit. As a result, a typical tomato harvest contains up to 60% immature green fruit (S. Sargent, personal communication). Immature green fruit are far inferior in flavour even compared to the ripened mature green product.

An ideal fresh tomato would produce fruit which can be allowed to remain on the vine long enough to develop full flavour while still maintaining the appropriate firmness and shelf life properties to permit cost effective production of the tomato. We reasoned that, to accomplish this, the process of fruit maturation must be delayed.

The phytohormone ethylene is known to regulate tomato fruit development (Reid 1987). Increases in the level of ethylene correlate with tomato fruit ripening and inhibitors of ethylene synthesis or perception delay fruit ripening. Ethylene is synthesized from methionine via the intermediate 1-aminocyclopropane-1-carboxylate (ACC; McKeon & Yang 1987). We proposed to reduce levels of ethylene and thereby delay fruit development through the selective degradation of ACC. Previous studies have identified a *Pseudomonas* enzyme, ACC deaminase, which selectively degrades ACC to α-ketobutyrate and ammonia (Honma & Shimomura 1978). To obtain a similar enzyme a screening program was established to identify microorganisms which were capable of utilizing ACC as a sole nitrogen source. Several bacterium were identified using this procedure and the gene encoding the enzyme responsible for the ACC degradation was subcloned. Biochemical analyses showed this enzyme to be ACC deaminase (Klee *et al.* 1991).

The ACC deaminase gene was engineered for plant expression using the cauliflower mosaic virus 35*S* promoter (Klee *et al.* 1991). This promoter will direct expression of the ACC deaminase protein essentially throughout the whole plant. Greenhouse grown, transgenic tomato plants were evaluated to determine the level of ethylene in leaves. Several lines were identified which contained significantly lower levels of ethylene. The ethylene level in leaves of tomato line 5673 was only 5% of that present in the leaves of non-transformed control plants (Klee *et al.* 1991). No obvious phenotypic effect was observed during the development of the ACC deaminase transgenic plants.

Transgenic and control tomato fruit showed similar profiles for ethylene synthesis throughout fruit development, however, the absolute levels of ethylene in the transgenic fruit were 90% lower than observed for the control (Klee *et al.* 1991). The time to ripen while on the vine was not affected at this level of ethylene synthesis. However, there was a significant reduction in overripening for transgenic fruit left on the vine. When fruit were picked at the breaker stage (first sign of colour change) the transgenic fruit showed significant increases in the length of time to develop full red colour, as compared to non-transformed controls, and in some cases transgenic fruit never developed full red colour. In addition, these transgenic fruit maintained firmness, 'on the shelf', for an extended period of time.

We have demonstrated that expressing ACC deaminase in tomato plants significantly reduces the level of ethylene. Transgenic tomato fruit, picked at a breaker stage, show a delay in time to red and a delay in overripening. We believe the ACC deaminase technology will serve as a powerful tool in modulating tomato fruit ripening and allow the development of fresh tomato varieties with both ideal flavour and handling attributes.

2. INCREASED STARCH SYNTHESIS IN POTATO

Potato represents a major staple food in many parts of the world, providing a rich source of dietary carbohydrate. The total market for potato grown in the U.S. alone is well over $2 billion annually, with the value determined predominantly by the food processing industry. The high value processing qualities include dry matter content, cold storage properties, disease tolerance, and chipping and french fry qualities. The most important determinant for these qualities is the dry matter content. A high solids potato processes better due to lower water content (which also reduces transportation and cooking costs), leading to increased processed-product recovery and reduced oil absorption during cooking that results in a product with reduced calories.

The major contributor to dry matter content is starch (Bajaj 1987), comprising about 70% of the dry matter of the mature tuber. Breeding efforts have been unsuccessful in making large gains in starch content in commercial cultivars, mostly because of the recalcitrant nature of potato breeding due to tetrasomic inheritance of characters, male sterility and self-incompatibility in many cultivars, and to high heterozygosity (Wenzel 1980). However, potato has proved amenable to manipulation through transgenic technology (Kaniewski *et al.* 1990; Rocha-Rosa *et al.* 1989), and this could provide a route for increasing starch content and improving dry matter yield in transgenic tubers.

Starch biosynthesis in plants (and glycogen biosynthesis in bacteria) requires the participation of the enzymes ADPglucose pyrophosphorylase (ADPGPP), starch (or glycogen) synthase, and branching enzyme (Preiss 1988, 1991). The first enzyme in the pathway, ADPGPP, utilizes glucose-1-phosphate and ATP to form ADPglucose, which subsequently serves as the substrate for starch synthase. The linear α-1,4-glucan polymer formed by the synthase may be subsequently branched by the activity of branching enzyme(s). In view of its sensitivity to allosteric effectors, ADPGPP has been suggested to be pivotal in plant starch biosynthesis, as it is in the bacterial pathway for glycogen biosynthesis. Therefore it was of interest to determine if enhanced ADPGPP activity would lead to increased starch biosynthesis in plants.

Plant ADPGPPs are tetramers that contain two distinct subunits and are regulated by 3-phosphoglyceric acid (PGA) and P_i as positive and negative effectors, respectively (Preiss 1988, 1991). Although both large and small subunits (Anderson *et al.* 1989; Olive *et al.* 1989; Bhave *et al.* 1990; Muller *et al.* 1990) show homology to the *E. coli* ADPGPP enzyme, the individual subunits do not appear to be catalytically active as evidenced by the absence of significant enzyme activity of the maize and *Arabidopsis* ADPGPP mutants (Bajaj 1987; Lin *et al.* 1988). In contrast, the *E. coli* ADPGPP, encoded by the *glgC* gene (Okita *et al.* 1981), is a regulated homotetrameric enzyme which is activated by fructose 1,6-bisphosphate (FBP) (Preiss *et al.* 1966). FBP increases the catalytic activity (V_{max}), reduces the K_m for the substrates, and modulates the sensitivity of the enzyme to the inhibitors 5′-AMP and inorganic phosphate (P_i) (Preiss *et al.* 1966; Gentner & Preiss 1968). A mutant *E. coli* K12 strain, 618, accumulates approximately 33% higher quantities of glycogen than its wild-type parent due to an alteration in the regulatory properties of ADPGPP (Cattaneo *et al.* 1969; Creuzet-Sigal *et al.* 1972). The mutant enzyme is less dependent on the activator, FBP, and is less sensitive to inhibition by the inhibitor, 5′-AMP, and this is due to a substitution of aspartic acid for glycine at position 336 in the ADPGPP enzyme (Leung *et al.* 1986; Lee *et al.* 1987). In view of the difficulties involved in coordinating the expression of two distinct genes, we chose to express in transgenic plants an *E. coli* ADPGPP enzyme to probe the regulation of metabolic flux by ADPGPP. Additionally, to minimize interference by the complex allosteric regulation of the wild-type *E. coli* gene, we used the mutant *glgC16* gene.

We introduced the *glgC16* gene fused with a transit peptide coding region (CTP) into Russet Burbank potato plants under the control of a tuber-specific patatin promoter (Bevan *et al.* 1986). The use of the transit peptide was required to direct the enzyme into the plastid, which is the site of starch biosynthesis. We recovered numerous transgenic plants which were vegetatively identical to the Russet Burbank parent. Subsequent analysis of transgenic plants showed that tubers expressing the patatin-CTP-*glgC16* gene contain, on average, 35% more starch than control tubers (Stark *et al.* 1992). Some transgenic potato lines produced tubers with nearly 60% more starch than controls. Therefore, tissue-specific expression of the CTP-*glgC16* gene using a patatin promoter results in significant increases in starch content without adverse effects on plant growth or development. Similar results recently were obtained in a small scale field trial (Stark *et al.* 1991), indicating that ADPGPP activity is rate-limiting for starch biosynthesis under conditions of potato cultivation.

Although the presence of the CTP-*glgC16* gene product was absolutely required for increased starch production, the degree of increase in starch content did not totally correlate with the level of expression of GlgC16 in all potato tubers (Stark *et al.* 1992). This suggests that relatively low levels of expression of the *E. coli* ADPGPP gene are sufficient to overcome the limitation of ADPglucose availability for starch biosynthesis. Whether the poor correlation with expression levels is due to substrate availability for the ADPGPP reaction, or to an unrelated reaction in the starch biosynthetic pathway remains to be established.

Our results show that ADPGPP activity is rate-limiting in starch biosynthesis. However, previous studies have shown that the activity of starch synthase is only one-twentieth that of ADPGPP and the branching enzymes (Preiss 1988, 1991; Robinson & Preiss 1987). These results imply that starch synthase is the rate-limiting reaction in starch biosynthesis. In view of our finding that enhanced ADPGPP activity increases the starch content of plant cells, we reasoned that allosteric regulation may play a significant role in modulating ADPGPP activity under *in vivo* conditions. To evaluate the importance of allosteric regulation in the control of ADPGPP activity and thus starch content, we expressed the wild-type *E. coli* ADPGPP gene in potato tubers in a manner similar to that of the *glgC16* gene. Unlike the GlgC16 enzyme, wild-type GlgC is fully subject to allosteric regulation (Okita *et al.* 1981), although the primary effector molecules differ between the plant and bacterial enzymes. Thus, depending upon the relative concentrations of these effectors *in vivo*, expression of the CTP-*glgC* gene may also result in increased ADPGPP activity and starch content in a plant cell environment. However, analysis of potato tubers showed that expression of CTP-*glgC* did not result in a noticeable increase in starch content, even though expression levels were equivalent to those in tubers expressing CTP-*glgC16* (Stark *et al.* 1992). We conclude that CTP-*glgC* expression may lead to a slight, but not significant, level of starch increase in plant tissues. The kinetic characteristics of the GlgC and GlgC16 enzymes have been previously reported (Cattaneo *et al.* 1969; Creuzet-Sigal *et al.* 1972; Leung *et al.* 1986; Lee *et al.* 1987), and differ primarily in allosteric regulation. The V_{max} of each enzyme is the same under fully activated conditions. Together, these results indicate that ADPGPP activity is rate-limiting in starch biosynthesis, and it is the regulatory properties of ADPGPP that make it rate-limiting and not the amount of enzyme protein.

The increased starch content in tubers expressing *E. coli* ADPGPP not only supports the role of ADPGPP activity in tuber sink modulation, as previously postulated (Rocha-Rosa *et al.* 1989; Preiss 1991; Lin *et al.* 1988), but presents evidence that modulation of source–sink relationships in the direction of increasing sink strength may prove to be an effective and powerful method of increasing dry matter content in storage organs of many different crops.

3. INSECT CONTROL IN COTTON

Cotton is an important, high value crop. An estimated 85 million acres are grown worldwide with over 12 million acres grown in the United States annually. Insect damage is a serious problem in the intensively managed cotton growing areas of the U.S. causing an estimated $645 million dollars per year in yield losses and insect control costs (Suguiyama *et al.* 1988). The lepidopteran insects, particularly *Helicoverpa zea* (cotton bollworm) and *Heliothis virescens* (tobacco budworm) are responsible for a large amount of insect damage on the high value acres of the Mississippi Delta Region.

Chemical insecticides, particularly pyrethroids, have been an effective means of controlling these insects. This control can be expensive in years of heavy infestation such as 1992, requiring scouting of the fields, timing of applications, and escalating rates of application because of decreasing effectiveness. Biological insecticide alternatives such as *Bacillus thuringiensis,* the bacterium that produces an insect control protein which does have activity against these pests (Aronson *et al.* 1986) has not been adopted for widespread use. These treatments are expensive, have a relatively short duration of protection and are not targeted to protect the productive parts of the plants such as the squares and bolls from insect damage.

Genes encoding the insect control proteins of *Bacillus thuringiensis* have been cloned, characterized and expressed in plants by several groups (Vaeck *et al.* 1987; Fischhoff *et al.* 1987). These plants exhibited insect control in the most sensitive lepidopteran insects but were limited by their low levels of expression of the insect control protein. Modified insect control protein genes, encoding proteins identical in amino acid sequence to the native genes of *Bacillus thuringiensis* have been more efficiently expressed in plants (Perlak *et al.* 1990, 1991). Among the factors dictating the modifications in the coding sequence of these genes are multiple sequence motifs that are not common in the coding regions of plant genes. These include localized regions of A+T nucleotide richness resembling plant introns, potential plant polyadenylylation signal sequences, ATTTA sequences which have been shown to destabilize mRNA in other systems and codons rarely used in plants.

The levels of expression of these modified genes in cotton (Perlak *et al.* 1990) have proven to be adequate first in greenhouse assays and then in the field tests that have been reported (Wilson *et al.* 1992) to control key economically important lepidopteran insects such as the tobacco budworm, cotton bollworm and pink bollworm (*Pectinophora gossypiella*). Extensive field test experimentation is currently underway to define the levels of control that these plants can offer the cotton grower and preliminary results are very encouraging. The success of these plants in controlling economically important cotton pests will offer the cotton grower an increased number of options to address his pest problems through more flexible integrated pest management programs. It could result in reduced production costs, reduced grower and environmental exposure to chemical insecticides and increase the effectiveness of the chemical applications that are utilized to manage crop production. This product will require a change in some agricultural practices in growing cotton and innovative crop management to recognize the full value of cotton protected from lepidopteran insect attack. These steps will make growers more productive and efficient in continuing to provide products of value to the world.

4. CONCLUSIONS

The above examples of transgenic crops indicate how new traits can improve the way crops are grown and utilized as well as improving their quality. The use of cotton plants that are naturally resistant to insect feeding reduces the amount of insecticides needed to grow the crop. Tomatoes with delayed ripening characteristics reduce the amount of spoilage after harvest and might provide a better tasting tomato to the grocery store. High starch potatoes might provide a healthful benefit by absorbing less oil during frying, resulting in less fat and less calories in the french fry or potato chip. These types of improved crop traits are only the beginning of the advances modern genetic engineering will bring to agriculture.

REFERENCES

Anderson, J.M., Okita, T.W. & Preiss, J. 1990 Enhancing carbon flow into starch: the role of ADPglucose pyrophosphorylase. In *The molecular and cellular biology of the potato* (ed. M. E. Vayda & W. D. Park), pp. 159–180. Wallingford: C.A.B. International.

Aronson, A.I., Beckman, W. & Dunn, P. 1986 *Bacillus thuringiensis* and related insect pathogens. *Microbiol. Rev.* **50**, 1–24.

Bajaj, Y.P.S. 1987 Biotechnology and the 21st century potato. In *Biotechnology in agriculture and forestry*, vol. 3. (ed. Y. P. S. Bajaj), pp. 3–22. Berlin and Heidelberg: Springer-Verlag.

Bevan, M., Barker, R., Goldsbrough, A., Jarvis, M., Kavanagh, T. & Iturriaga, G. 1986 The structure and transcription start site of a major tuber protein gene. *Nucl. Acids Res.* **14**, 4625.

Bhave, M.R., Lawrence, S., Barton, C. & Hannah, L.C. 1990 Identification and molecular characterization of *Shrunken-2* cDNA clones of maize. *Pl. Cell* **2**, 581.

Cattaneo, J., Damotte, M, Sigal, N., Sanchez-Medina, F. & Puig, J. 1969 Genetic studies of *Escherichia coli* K12 mutants with alterations in glycogenesis and properties of an altered adenosine-diphosphate-glucose-pyrophosphorylase. *Biochem. biophys. Res. Commun.* **34**, 694–701.

Creuzet-Sigal, N., Latil-Damotte, M. & Puig, J. 1972 Genetic analysis and biochemical characterization of mutants impairing glycogen metabolism in *Escherichia coli* k12. In *Biochemistry of the glycosidic linkage* (ed. R. Piras & H. G. Pontis), vol. 2, pp. 647–680. The Pan-American Association of Biochemical Societies Symposium. New York: Academic Press.

Gentner, N. & Preiss, J. 1968 Biosynthesis of bacterial glycogen. VI. Differences in the kinetic properties of the *Escherichia coli* B adenosine diphosphate glucose pyrophosphorylase depending on whether Mg+2 or Mn+2 serves as divalent cation. *J. biol. Chem.* **243**, 5882.

Honma, M. & Shimomura, T. 1978 Metabolism of 1-amino-cyclopropane-1-carboxylic Acid. *Agric. biol. Chem.* **42**, 1825–1831.

Kader, A., Stevens, M.A., Albright-Holton, M., Morris, L. & Algazi, M. 1977 Effect of fruit ripeness when picked on flavor and composition in fresh market tomatoes. *J. Am. Soc. Hort. Sci.* **102**, 724–731.

Kaniewski, W., Lawson, C., Sammons,B. *et al.* 1990 Field resistance of transgenic Russet Burbank potato to effects of infection by potato virus X and potato virus Y. *Bio Technology* **8**, 750–754.

Klee, H., Hayford, M., Kretzmer, K., Barry, G. & Kishore, G. 1991 Control of ethylene synthesis by expression of a bacterial enzyme in transgenic tomato plants. *Pl. Cell* **3**, 1187–1193.

Lee, Y.M., Kumar, A. & Preiss, J. 1987 Amino acid sequence of an *Escherichia coli* ADPglucose synthetase allosteric mutant as deduced from the DNA sequence of the *glgC* gene. *Nucl. Acids Res.* **15**, 10603.

Leung, P., Lee, Y.-M., Greenberg, E., Esch, K., Boylan, S. & Preiss, J. 1986 Cloning and expression of the *Escherichia coli glgC* gene from a mutant containing an ADPglucose pyrophosphorylase with altered allosteric properties. *J. Bacteriol.* **167**, 82–88.

Lin, T.P., Caspar, T., Somerville, C. & Preiss, J. 1988 Isolation and characterization of a starchless mutant of *Arabidopsis thaliana* (L) Henyh lacking ADPglucose pyrophosphorylase activity. *Pl. Physiol.* **86**, 1131–1135.

McKeon, T. & Yang, S.F. 1987 Biosynthesis and metabolism of ethylene. In *Plant hormones and their role in plant growth and development* (ed. P. J. Davies), pp. 94–112. Martinus Nijhoff Publishers/Kluwer Academic.

Muller, B.T., Koßmann, J., Hannah, L.C., Willmitzer, L. & Sonnewald, U. 1990 One of two different ADP-glucose pyrophosphorylase genes from potato responds strongly to elevated levels of sucrose. *Molec. Gen. Genet.* **224**, 136–146.

Okita, T.W., Rodriguez, R.L. & Preiss, J. 1981 Biosynthesis of bacterial glycogen: cloning of the glycogen biosynthetic enzyme structural genes of *Escherichia coli. J. biol. Chem.* **256**, 6944.

Olive, M.R., Ellis, R.J. & Schuch, W.W. 1989 Isolation and nucleotide sequences of cDNA clones encoding ADP-glucose pyrophosphorylase peptides from wheat leaf and endosperm. *Pl. molec. Biol.* **12**, 525.

Preiss, J. 1988 Biosynthesis of starch and its regulation. In *The biochemistry of plants* (ed. J. Preiss), vol. 14, pp. 181–254. San Diego: Academic Press.

Preiss, J. 1991 Biology and molecular biology of starch synthesis and its regulation. In *Oxford surveys of plant molecular and cell biology*, vol. 7 (ed. B. Miflin), pp. 59–114. Oxford University Press.

Preiss, J., Shen, L., Greenberg, E. & Genter, N. 1966 Biosynthesis of bacterial glycogen. IV. Activation and inhibition of the adenosine diphosphate glucose pyrophosphorylase of *Escherichia coli* B. *Biochemistry* **5**, 1833.

Reid, M. 1987 Ethylene in plant growth, development and senescence. In *Plant hormones and their role in plant growth and development* (ed. P. J. Davies), pp. 257–279. Martinus Nijhoff Publishers/Kluwer Academic.

Robinson, N.L. & Preiss, J. 1987 Localization of carbohydrate metabolizing enzymes in guard cells of *Commelina communis*. *Pl. Physiol.* **85**, 360.

Rocha-Rosa, M., Sonnewald, U., Frommer, W., Stratmann, M., Schell, J. & Willmitzer, L. 1989 Both developmental and metabolic signals activate the promoter of a class I patatin gene. *EMBO J.* **8**, 23–31.

Stark, D.M., Barry, G.F., Muskopf, Y.M., Timmerman, K.P., Kishore, G.M. 1991 Increased starch and dry matter deposition in transgenic russet burbank potato tubers. Paper presented at the Third International Congress of Plant Molecular Biology, Tucson, Arizona, 6–11 October 1991, abstract 714.

Stark, D.M., Timmerman, K.P., Barry, G.F., Preiss, J. & Kishore, G.M. 1992 Regulation of the amount of starch in plant tissues by ADPglucose pyrophosphorylase. *Science, Wash.* **258**, 287.

Suguiyama, L. & Osteen, C. 1988 The economic importance of cotton insects and mites. United States Department of Agriculture Economic Research Service, Agricultural Economic Report Number 599.

Vaeck, M., Reynaerts, A., Hofte, H. *et al.* 1987 Transgenic plants protected from insect attack. *Nature, Lond.* **328**, 33–37.

Von Scheele, C. 1937 Die Bestimmung des Starkghelts und der Trockensubstanz der Kartoffel mit hilfe des Specifischen gewichte.*Landw Ver Stn* **127**, 67–96.

Wenzel, G. 1980 The potentials and limits of classical genetics in plant breeding. In *Plant cell cultures: results and perspectives* (ed. F. Sala, B. Paris, R. Cella & O. Cifferri), pp. 33–47. Elsevier.

Wilson, F.D., Flint, H.M., Deaton, W.R. *et al.* 1992 Resistance of cotton lines containing a *Bacillus thuringiensis* toxin to Pink Bollworm (Lepidoptera: Gelechiidae) and other insects. *J. econ. Entomol.* **85**, 1516–1521.

13

Transgenic investigations of prion diseases of humans and animals

S. B. Prusiner

SUMMARY

Prions cause transmissible and genetic neurodegenerative diseases. Infectious prion particles are composed largely, if not entirely, of an abnormal isoform of the prion protein (PrP^{Sc}), which is encoded by a chromosomal gene. Although the PrP gene is single copy, transgenic mice with both alleles of the PrP gene ablated develop normally. A post-translational process, as yet unidentified, converts the cellular prion protein (PrP^{C}) into PrP^{Sc}. Scrapie incubation times, neuropathology and prion synthesis in transgenic mice are controlled by the PrP gene. Mutations in the PrP gene are genetically linked to development of neurodegeneration. Transgenic mice expressing mutant PrP spontaneously develop neurological dysfunction and spongiform neuropathology. Investigations of prion diseases using transgenesis promise to yield much new information about these once enigmatic disorders.

1. INTRODUCTION

The prion diseases are a group of neurodegenerative disorders of animals and humans. These diseases are transmissible under some circumstances to experimental animals by inoculation. Unlike other transmissible disorders, the prion diseases can also be caused by mutations in the prion protein, PrP, which is encoded by a chromosomal gene. Four diseases of animals and four of humans are caused by prions (table 1). Scrapie of sheep and goats is the prototypic prion disease. Mink encephalopathy, chronic wasting disease and bovine spongiform encephalopathy (BSE) are all thought to occur after the consumption of prion-infected foodstuffs. Similarly, kuru of the New Guinea Fore people is thought to have resulted from the consumption of brains from dying relatives during ritualistic cannibalism (Alpers 1979; Gajdusek 1977). Creutzfeldt–Jakob disease (CJD) occurs primarily as a sporadic disorder (Masters *et al.* 1981*b*) but iatrogenic CJD is thought to result from the accidental inoculation of patients with prions (Fradkin *et al.* 1991; Gibbs *et al.* 1985). Familial CJD, Gerstmann–Sträussler–Scheinker syndrome (GSS) and fatal familial insomnia are all dominantly inherited prion diseases which have been shown to be caused by mutations in the PrP gene (Brown *et al.* 1991; Collinge *et al.* 1989; Hsiao & Prusiner 1990; Medori *et al.* 1992*b*).

For more than a century, scrapie was considered an enigmatic disorder of sheep and goats, the etiology of which was unknown (M'Gowan 1914; Parry 1983). By 1938, experimental transfer of scrapie from one sheep to another began to suggest an infectious etiology (Cuillé & Chelle 1939). Meanwhile, observations that the genetic backgrounds of flocks profoundly influence their susceptibility to scrapie raised the possibility that scapie might be an inherited disorder (Gordon 1966). These opposing views sparked many controversial encounters (Dickinson *et al.* 1965; Parry 1962) and foreshadowed a series of equally bitter arguments about the possible structure of the transmissible scrapie agent (Pattison 1988).

Over the past decade, a growing body of experimental data has begun to provide a coherent yet unprecedented picture of the novel infectious pathogens or prions causing scrapie (Prusiner 1982, 1991). Whereas inherited, transmissible and sporadic prion diseases of humans are now well documented, the situation with natural prion diseases of animals is less clear. Progress in understanding the human prion diseases has its roots in their transmission to animals (Masters *et al.* 1979, 1981*a*) and the discovery of the prion protein (PrP) (Bolton *et al.* 1982; Prusiner 1982) followed by the molecular cloning of the PrP gene (Chesebro *et al.*, 1985; Oesch *et al.*, 1985; Prusiner *et al.* 1984).

As molecular, biological and genetic analyses of both the human and animal prion diseases have advanced, the biochemistry of the prion protein has continued to pose both methodological and conceptual problems. For example, transmissible prions are composed largely, if not entirely, of an abnormal isoform of cellular PrP designated PrP^{Sc} (Gabizon & Prusiner 1990; Prusiner 1991). Although PrP^{Sc} is synthesized from cellular PrP (PrP^{C}) by a post-translational process (Basler *et al.* 1986; Borchelt *et al.* 1990, 1992; Caughey & Raymond 1991), the precise nature of this protein transformation remains unknown. Whether the conversion of PrP^{C} to PrP^{Sc} involves an as yet unidentified chemical modification,

Table 1. *Prion diseases*[1]

disease	natural host
scrapie	sheep and goats
transmissible mink encephalopathy (TME)	mink
chronic wasting disease (CWD)	mule deer and elk
bovine spongiform encephalopathy (BSE)	cattle
kuru	humans – fore[2]
Creutzfeldt–Jakob disease (CJD)	humans
Gerstmann–Sträussler–Scheinker syndrome (GSS)	humans
fatal familial insomnia (FFI)	humans

[1] Alternative terminologies include slow virus infections, subacute transmissible spongiform encephalopathies, and unconventional slow virus diseases (Gajdusek 1977).

[2] Kuru is confined to the Fore tribe and surrounding tribes in the highlands of Papua New Guinea.

perhaps labile under the conditions of analysis, or whether it only involves a conformational change (Stahl *et al.* 1992b) remains to be established.

To date, it has not been possible to synthesize PrP^{Sc} in cell-free systems (Raeber *et al.* 1992), but studies of this insoluble protein in cultured cells have yielded information about the subcellular site of its synthesis and deposition (Borchelt *et al.* 1992; McKinley *et al.* 1991*b*). In the brains of animals and humans dying of prion diseases, PrP^{Sc} is found in the neuropil (Taraboulos *et al.* 1992*a*) and sometimes in the extracellular space as discrete accumulations called plaques (DeArmond *et al.* 1985; Kitamoto *et al.* 1986). These PrP plaques were first described as amyloid deposits because they exhibited a green-gold birefringence after staining with Congo red dye when viewed by polarization microscopy (Kiatzo *et al.* 1959). When present, PrP amyloid plaques are diagnostic of prion diseases. Rod-shaped polymers of PrP with the properties of amyloid can be generated by limited protease digestion of PrP^{Sc} in the presence of detergent (McKinley *et al.* 1991*a*; Prusiner *et al.* 1983).

The function of PrP^{C} is unknown but PrP^{C} molecules appear to be unnecessary because mice homologous for disruption of the PrP gene develop normally and are healthy for more than 9 months (Büeler *et al.* 1992). These results argue that scrapie and the other prion diseases do not result from an inhibition of PrP^{C} function caused by PrP^{Sc}, but rather the accumulation of PrP^{Sc} interferes with some as yet undefined cellular process.

2. THE PRION PROTEIN

Once it was established that scrapie prion infectivity depended upon protein (Prusiner *et al.* 1981), the search for a scrapie-specific protein intensified. Although the insolubility of scrapie infectivity made purification problematic, we took advantage of this property, along with its relative resistance to degradation by proteases, to extend the degree of purification. Radio-iodination of partly purified fractions revealed a protein unique to preparations from scrapie-infected brains (Bolton *et al.* 1982; Prusiner *et al.* 1982). This protein was later named prion protein (PrP) with an apparent molecular mass of 27–30 kDa, or PrP 27–30 (McKinley *et al.* 1983*a*).

Subsequent studies showed that PrP 27–30 is derived from a larger protein of molecular mass 33–35 kDa, designated PrP^{Sc} (Meyer *et al.* 1986; Oesch *et al.* 1985). At the same time it was found that the brains of normal and scrapie-infected hamsters express similar levels of PrP mRNA and a protease-sensitive prion protein, designated PrP^{C} (Oesch *et al.* 1985). The function of PrP^{C} is unknown, although it has been suggested that a PrP-like molecule from chickens may have acetylcholine receptor-inducing activity (Harris *et al.* 1991). Furthermore, PrP^{C} does not seem to be essential, at least in young mice, as disruption of the PrP gene has not caused any detectable abnormalities in the nervous, musculoskeletal or lymphoreticular systems at 9 months of age (Büeler *et al.* 1992). Perhaps the absence of PrP^{C} will result in abnormalities later in life, as is the case for the p53 tumor suppressor protein where young animals lacking p53 are normal but as they age neoplasms develop (Donehower *et al.* 1992).

3. STRUCTURE, ORGANIZATION AND EXPRESSION OF THE PrP GENE

The entire open reading frame (ORF) of all known mammalian and avian PrP genes is contained within a single exon (figure 1) (Basler *et al.* 1986; Gabriel *et al.* 1992; Hsiao *et al.* 1989*a*; Puckett *et al.* 1991; Westaway *et al.* 1989, 1991). This feature of the PrP gene eliminates the possibility that PrP^{Sc} arises from

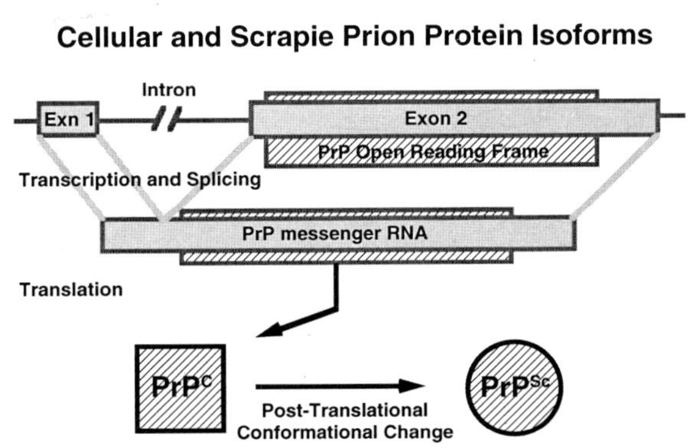

Figure 1. Structure and organization of the chromosomal prion protein gene. In all mammals examined, the entire open reading frame (ORF) is contained within a single exon. The 5′ untranslated region of the PrP mRNA is derived from either one or two additional exons (Basler *et al.* 1986; Puckett *et al.* 1991; Westaway *et al.* submitted. Only one PrP mRNA has been detected. PrP^{Sc} is thought to be derived from PrP^{C} by a post-translational process (Basler *et al.* 1986; Borchelt *et al.* 1990, 1992; Caughey & Raymond 1991; Taraboulos *et al.* 1992*b*). The amino acid sequence of PrP^{Sc} is identical to that predicted from the translated sequence of the DNA encoding the PrP gene (Basler *et al.* 1986; Stahl *et al.* 1992*b*) and no unique post-translational chemical modifications have been identified that might distinguish PrP^{Sc} from PrP^{C}. Thus, it seems likely that PrP^{C} undergoes a conformational change as it is converted to PrP^{Sc}.

alternative RNA splicing (Basler *et al.* 1986; Westaway *et al.* 1987, 1991); however, mechanisms such as RNA editing or protein splicing remain a possibility (Blum *et al.* 1990; Kane *et al.* 1990). The two exons of the Syrian hamster (SHa) PrP gene are separated by a 10 kilobase (kb) intron: exon 1 encodes a portion of the 5′ untranslated leader sequence, whereas exon 2 encodes the ORF and 3′ untranslated region (Basler *et al.* 1986). The mouse (Mo) PrP gene is comprised of three exons, with exon 3 analogous to exon 2 of the hamster (Westaway *et al.* 1991). The promoters of both the SHa and MoPrP genes contain copies of G–C rich repeats 3 and 2, respectively, but are devoid of TATA boxes. These G–C nonamers represent a motif which may function as a canonical binding site for the transcription factor Sp1 (McKnight & Tjian 1986).

Although PrP mRNA is constitutively expressed in the brains of adult animals (Chesebro *et al.* 1985; Oesch *et al.* 1985), it is highly regulated during development. In the septum, levels of PrP mRNA and choline acetyltransferase were found to increase in parallel during development (Mobley *et al.* 1988). In other brain regions, PrP gene expression occurred at an earlier age. *In situ* hybridization studies show that the highest levels of PrP mRNA are found in neurons (Kretzschmar *et al.* 1986*a*).

PrP^{C} expression in brain was defined by standard immunohistochemistry (DeArmond *et al.* 1987) and by histoblotting (Taraboulos *et al.* 1992*a*) (figure 2). Immunostaining of PrP^{C} in the SHa brain was most

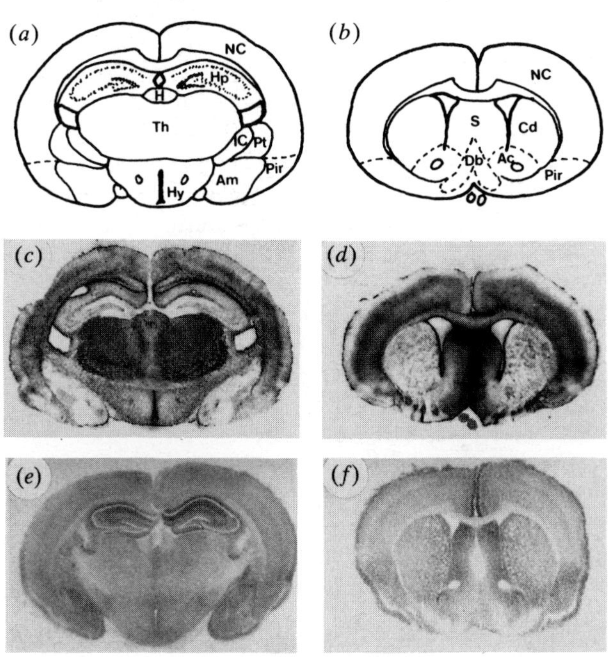

Figure 2. Histoblots of Syrian hamster brain immunostained for PrP^{C} or PrP^{Sc}. Coronal sections through the hippocampus-thalamus (*a,c,e*) and the septum-caudate (*b,d,f*). Brain sections of a Syrian hamster (*c,d*) clinically ill after inoculation with Sc237 prions and (*e,f*) an uninfected, control animal. Immunostaining for (*c,d*) PrP^{Sc} and (*e,f*) PrP^{C}. Ac, nucleus accumbens; Am, amygdala; Cd, caudate nucleus; Db, diagonal band of Broca; H, habenula; Hp, hippocampus; Hy, hypothalmus; IC, internal capsule; NC, neocortex; Th, thalamus; Pir, piriform cortex; Pt, putamen; S, septal nuclei. Reproduced from Taraboulos *et al.* (*Proc. natn. Acad. Sci. U.S.A.*, 1992*a*).

intense in the stratum radiatum and stratum oriens of the CA1 region of the hippocampus and was virtually absent from the granule cell layer of the dentate gyrus and the pyramidal cell layer throughout Ammon's horn. PrP^{Sc} staining was minimal in these regions which were intensely stained for PrP^{C}. A similar relation between PrP^{C} and PrP^{Sc} was found in the amygdala. In contrast, PrP^{Sc} accumulated in the medial habenular nucleus, the medial septal nuclei and the diagonal band of Broca; these areas were virtually devoid of PrP^{C}. In the white matter, bundles of myelinated axons contained PrP^{Sc} but were devoid of PrP^{C}. These findings suggest that prions are transported along axons in agreement with earlier findings where scrapie infectivity was found to migrate in a pattern consistent with retrograde transport (Fraser & Dickinson 1985; Jendroska *et al.* 1991; Kimberlin *et al.* 1983). Although the rate of PrP^{Sc} synthesis appears to be a function of the level of PrP^{C} expression in transgenic (Tg) mice, the level to which PrP^{Sc} accumulates appears to be independent of PrP^{C} concentration (Prusiner *et al.* 1990).

4. POST-TRANSLATIONAL SYNTHESIS OF PrP^{Sc}

Metabolic labelling studies of scrapie-infected cultured cells have shown that PrP^{C} is synthesized and degraded rapidly whereas PrP^{Sc} is synthesized slowly by an as yet undefined post-translational process (figure 1) (Borchelt *et al.* 1990, 1992; Caughey *et al.* 1989; Caughey & Raymond, 1991). These observations are consistent with earlier findings showing that PrP^{Sc} accumulates in the brains of scrapie-infected animals while PrP mRNA levels remain unchanged (Oesch *et al.* 1985). Furthermore, the structure and organization of the PrP gene made it likely that PrP^{Sc} is formed during a post-translational event (Basler *et al.* 1986).

Both PrP isoforms appear to transit through the Golgi apparatus where their Asn-linked oligosaccharides are modified and sialylated (Bolton *et al.* 1985; Endo *et al.* 1989; Haraguchi *et al.* 1989; Manuelidis *et al.* 1985; Rogers *et al.* 1990). PrP^{C} is presumably transported within secretory vesicles to the external cell surface where it is anchored by a glycosyl phosphatidylinositol (GPI) moiety (Baldwin *et al.* 1990; Safar *et al.* 1990; Stahl *et al.* 1987, 1990*a,b*). In contrast, PrP^{Sc} accumulates primarily within cells where it is deposited in cytoplasmic vesicles, many of which appear to be secondary lysosomes (Borchelt *et al.* 1992; Butler *et al.* 1988; Caughey *et al.* 1991; McKinley *et al.* 1991*b*; Taraboulos *et al.* 1992*b*; Taraboulos *et al.* 1990*b*).

Whether PrP^{C} is the substrate for PrP^{Sc} formation or whether a restricted subset of PrP molecules are precursors for PrP^{Sc} remains to be established. Several experimental results suggest that PrP molecules destined to become PrP^{Sc} exit to the cell surface, as does PrP^{C} (Stahl *et al.* 1987), before their conversion into PrP^{Sc} (Borchelt *et al.* 1992; Caughey & Raymond 1991; Taraboulos *et al.* 1992*b*). Interestingly, the GPI anchors of both PrP^{C} and PrP^{Sc}, which presumably feature in directing the subcellular trafficking of these molecules, are sialylated (Stahl *et al.* 1992*a*). It is unknown whether sialylation of the GPI anchor participates in some aspect of PrP^{Sc} formation.

Although most of the difference in mass of PrP 27–30 predicted from the amino acid sequence and that observed after post-translational modification is due to complex-type oligosaccharides, these sugar chains are not required for the synthesis of protease-resistant PrP in scrapie-infected cultured cells based on experiments with the Asn-linked glycosylation inhibitor tunicamycin and on site-directed mutagenesis studies (Taraboulos *et al.* 1990*a*). Whether unglycosylated PrP^{Sc} is associated with scrapie prion infectivity remains to be established, but experiments with transgenic mice may resolve this issue.

Cell-free translation studies have demonstrated two forms of PrP: a transmembrane form which spans the bilayer twice at the transmembrane (TM) and amphipathic helix (AH) domains, and a secretory form (Bazan *et al.* 1987; Hay *et al.* 1987*a,b*; Lopez *et al.* 1990; Yost *et al.* 1990). The stop transfer effector (STE) domain controls the topogenesis of PrP. That PrP contains both a TM domain and a GPI anchor poses a topological conundrum. It seems likely that membrane-dependent events feature in the synthesis of PrP^{Sc}, especially as brefeldin A, which selectively destroys the Golgi stacks (Doms *et al.* 1989; Lippincott-Schwartz *et al.* 1989), prevents PrP^{Sc} synthesis in scrapie-infected cultured cells (Taraboulos *et al.* 1992*b*). For many years, the association of scrapie infectivity with membrane fractions has been appreciated (Gibbons & Hunter 1967; Griffith 1967; Millson *et al.* 1971); indeed, hydrophobic interactions are thought to account for many of the physical properties displayed by infectious prion particles (Gabizon *et al.* 1987; Prusiner *et al.* 1978, 1980).

5. PRION DISEASES OF SHEEP AND CATTLE

Even though scrapie was recognized as a distinct disorder of sheep with respect to its clinical manifestations as early as 1738, the disease remained enigmatic even with respect to its pathology for more than two centuries. Some veterinarians thought that scrapie was a disease of muscle caused by parasites, whereas others thought that it was a dystrophic process. An investigation into the etiology of scrapie followed the vaccination of sheep for looping ill virus with formalin-treated extracts of ovine lymphoid tissue unknowingly contaminated with scrapie prions (Gordon 1946). Two years later, more than 1500 sheep developed scrapie from this vaccine.

While the transmissibility of scrapie became well established, the spread of scrapie within and among flocks of sheep remained puzzling. Parry argued that host genes were responsible for the development of scrapie in sheep. He was convinced that natural scrapie is a genetic disease which could be eradicated by proper breeding protocols (Parry 1962; Parry 1983). He considered its transmission by inoculation of importance primarily for laboratory studies and communicable infection of little consequence in nature. Scrapie is widely recognized as a naturally transmissible disease of sheep and goats, and it has been argued that host genetics only modulates susceptibility to an endemic infectious agent (Dickinson *et al.* 1965).

Studies of PrP genes (*Prn-p*) in mice have revealed that short or long incubation times occur before scrapie. A genetic linkage has been demonstrated between a *Prn-p* restriction fragment length polymorphism and a gene modulating incubation times (*Prn-i*) (Carlson *et al.* 1986). Other investigators have confirmed the genetic linkage, and one group has shown that the incubation time gene *Sinc* is also linked to PrP (Carlson *et al.* 1988; Hunter *et al.* 1987; Race *et al.* 1990). The incubation time gene for experimental scrapie in Cheviot sheep called *Sip* is said to be linked to a PrP gene restriction fragment length polymorphism (Hunter *et al.* 1989), a situation perhaps analogous to *Prn-i* and *Sinc* in mice. *Sinc* was first described by Dickinson and colleagues over 20 years ago (Dickinson *et al.* 1968); whether the genes for PrP, *Prn-i* and *Sinc* are all congruent remains to be established. The PrP sequences of NZW ($Prn\text{-}p^{a}$) and I/Ln ($Prn\text{-}p^{b}$) mice with short and long scrapie incubation times, respectively, differ at codons 108 (L→F) and 198 (T→V) (Westaway *et al.* 1987). Although these amino acid substitutions argue for the congruency of *Prn-p* and *Prn-i*, experiments with $Prn\text{-}p^{a}$ mice expressing $Prn\text{-}p^{b}$ transgenes demonstrated a paradoxical shortening of incubation times (Westaway *et al.* 1991) instead of a prolongation as predicted from ($Prn\text{-}p^{a} \times Prn\text{-}p^{b}$) F1 mice which exhibit long incubation times that are dominant (Carlson *et al.* 1986, 1988; Dickinson *et al.* 1968; Hunter *et al.* 1987; Race *et al.* 1990).

Since 1986 more than 70 000 cattle have been killed with BSE in Great Britain (Dealler & Lacey 1990; Wilesmith & Wells 1991; Wilesmith *et al.* 1988, 1992*a,b*). Neither the cause of BSE, often referred to as 'mad cow disease' nor methods of controlling the spread of this disorder are known. Many investigators contend that BSE resulted from the feeding of dietary protein supplements derived from rendered scrapie-infected sheep offal to cattle, a practice banned since 1988. Curiously, the majority of BSE cases have occurred in herds with a single affected animal within a herd; several cases of BSE in a single herd are infrequent (Dealler & Lacey 1990; Wilesmith & Wells 1991; Wilesmith *et al.* 1988). Whether the distribution of BSE cases within herds will change as the epidemic progresses and BSE will disappear with the cessation of feeding rendered meat and bone meal are uncertain.

6. HUMAN PRION DISEASES

The discovery of human prion diseases came from the recognition that the neuropathology of the cerebellar disorder kuru, which is confined to natives in the Fore region of New Guinea (Gajdusek 1977; Gajdusek *et al.* 1966), was similar to that of scrapie. Once the most common cause of death among women and children, kuru has almost disappeared with the cessation of ritualistic cannibalism (Alpers 1987). These findings suggest that kuru was transmitted orally. Of note are recent cases of kuru which have occurred in people exposed to prions more than three decades ago (Prusiner *et al.* 1982). Spongiform degeneration in kuru prompted Hadlow (1959) to suggest that transmission studies in apes be done. The success of those studies (Gajdusek *et al.* 1966) was followed by the transmission of CJD to apes (Gibbs *et al.* 1968) based on the earlier recognition that the neuropathological changes in kuru were similar to those found in CJD (Klatzo *et al.* 1959). In 1920, Creutzfeldt reported the case of a 23-year-old women who died of a neurodegenerative disease, and the following year Jakob reported five cases (Jakob 1921*a*,*b*,*c*). Ironically, some investigators doubt that Creutzfeldt described the disease that now bears his name (Richardson 1977).

In humans, a genetic basis of the condition was first thought to have a role in CJD with the recognition that ~10% of cases are familial (Gajdusek 1977; Masters *et al.* 1981*b*). Like sheep scrapie, the relative contributions of genetic and infectious etiologies in the human prion diseases remained puzzling. The discovery of the PrP gene raised the possibility that mutation might feature in the hereditary human prion diseases. A point mutation at codon 102 (P→L) was shown to be linked genetically to development of Gerstmann–Straussler–Scheinker disease (GSS) with a LOD score exceeding 3 (figure 3) (Hsiao *et al.* 1989*a*). This mutation may be caused by the deamination of a methylated CpG in a germline PrP gene resulting in the substitution of a T for C. The codon 102 mutation has been found in ten different families in nine different countries including the original GSS family (Doh-ura *et al.* 1989; Goldgaber *et al.* 1989; Kretzschmar *et al.* 1991*a*,*b*).

An insert of 144 base pairs (b.p.) at codon 53 containing six octarepeats has been described in patients with CJD from four families, all residing in southern England (Collinge *et al.* 1989, 1990; Crow *et al.* 1990; Owen *et al.* 1989, 1990, 1991). This mutation must have arisen through a complex series of events because the human PrP gene contains only five octarepeats, suggesting that a single recombination event could not have created the insert. Genealogic investigations have shown that all four families are related, suggesting a single founder born more than two centuries ago (Crow *et al.* 1990). The LOD score for this extended pedigree exceeds 11. Studies from several laboratories have demonstrated that four, five, six, seven, eight or nine octarepeats in addition to the normal five are found in individuals with inherited CJD, whereas deletion of one octarepeat has been identified without the neurologic disease (Collinge *et al.* 1989, 1990; Goldfarb *et al.* 1991*a*; Laplanche *et al.* 1990; Owen *et al.* 1989, 1990, 1992; Vnencak-Jones & Phillips 1992).

For many years, the unusually high incidence of CJD among Israeli Jews of Libyan origin was thought to be caused by the consumption of lightly cooked sheep brain or eyeballs (Alter & Kahana 1976; Herzberg *et al.* 1974; Kahana *et al.* 1974; Neugut *et al.* 1979). Recent studies have shown that some Libyan

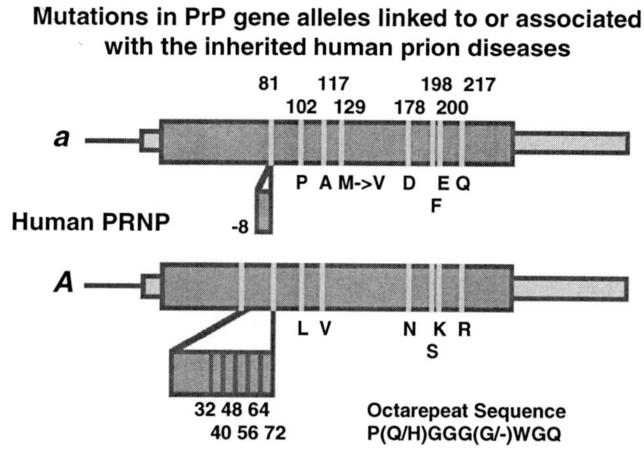

Figure 3. Human prion protein gene. The open reading frame (ORF) is denoted by the large grey rectangles and the exon by the smaller rectangles. Codon numbers are indicated above the amino acid sequence. Human PrP wild-type polymorphisms are shown in the upper rectangle denoted 'a' whereas mutations linked to or associated with prion diseases are depicted in the lower rectangle denoted 'A'. The wild-type human PrP gene contains five octarepeats [P(Q/H)GGG(G/−)WGQ] from codons 51 to 91 (Kretzschmar *et al.* 1986*b*). Deletion of a single octarepeat at codon 81 or 82 is not associated with prion disease (Laplanche *et al.* 1990; Puckett *et al.* 1991; Vnencak-Jones & Phillips 1992). Whether such a deletion alters the phenotypic characteristics of a prion disease is unknown, but homozygosity for Met or Val at codon 129 appears to increase susceptibility to sporadic CJD (Palmer *et al.* 1991). Octarepeat inserts of 32, 40, 48, 56, 64, and 72 amino acids at codons 67, 75 or 83 have been found and are either genetically linked to or associated with familial CJD (Collinge *et al.* 1989, 1990; Crow *et al.* 1990; Goldfarb *et al.* 1990*c*, 1991*a*; Owen *et al.* 1989, 1990; J. Collinge & M. S. Palmer, unpublished data). Point mutations at codons 102 (Pro→Leu), 117 (Ala→Val), and 198 (Phe→Ser) are found in patients with GSS (Doh-ura *et al.* 1989; Goldfarb *et al.* 1990*a*,*c*,*d*; Goldgaber *et al.* 1989; Hsiao *et al.* 1989*a*,*b*, 1991*b*; Hsiao & Prusiner 1990; Tateishi *et al.* 1990). There are common polymorphisms at codons 117 (Ala→Ala) and 129 (Met→Val). Point mutations at codons 178 (Asp→Asn) and 200 (Glu→Lys) are found in patients with familial CJD (Gabizon *et al.* 1991; Goldfarb *et al.* 1990*b*, 1991*c*; Hsiao *et al.* 1991*a*). Point mutations at codons 198 (Phe→Ser) and 217 (Gln→Arg) are found in patients with GSS who have PrP amyloid plaques and neurofibrillary tangles (Dlouhy *et al.* 1992; Hsiao *et al.* 1992). Single letter code for amino acids is as follows: A, Ala; D, Asp; E, Glu; F, Phe; K, Lys; L, Leu; M, Met; N, Asn; P, Pro; Q, Gln; R, Arg; S, Ser; T, Thr; and V, Val.

and Tunisian Jews in families with CJD have a PrP gene point mutation at codon 200 resulting in a E→K substitution (Gabizon *et al.* 1991; Goldfarb *et al.* 1990; Hsiao *et al.* 1991*a*). One patient was homozygous for the mutation, but her clinical presentation was similar to that of heterozygotes (Hsiao *et al.* 1991*a*), suggesting that familial prion diseases are true autosomal dominant disorders like Huntington's disease (Wexler *et al.* 1987). The codon 200 mutation has also been found in Slovaks originating from Orava in North Central Czechoslovakia (Goldfarb *et al.* 1990), in a cluster of familial cases in Chile (Goldfarb *et al.* 1991*b*) and in a large German family living in the United States (Bertoni *et al.* 1992). Some investigators have argued that the codon 200 mutation originated in a Sephardic Jew whose descendants migrated from Spain and Portugal at the time of the inquisition (Goldfarb *et al.* 1991*b*). It is more likely that the codon 200 mutation has arisen independently multiple times by the deamidation of a methylation CpG as described above the codon 102 mutation (Hsiao *et al.* 1989*a*, 1991*a*). In support of this hypothesis are historical records of Libyan and Tunisian Jews showing that they are descended from Jews living on the island of Jerba where Jews first settled around 500 BC and not from Sephardim (Udovitch & Valensi 1984).

Many families with CJD have been found to have point mutations at codon 178 (Brown *et al.* 1992; Fink *et al.* 1991; Goldfarb *et al.* 1991*c*, 1992; Haltia *et al.* 1991). In these patients, as well as those with the codon 200 mutation, PrP amyloid plaques are rare; the neuropathological changes generally consist of widespread spongiform degeneration. Recently, a new prion disease which presents with insomnia has been described in three Italian families with the codon 178 mutation (Medori *et al.* 1992*a*,*b*). The neuropathology in these patients with fatal familial insomnia is restricted to selected nuclei of the thalamus. It is unclear whether all patients with the codon 178 mutation or only a subset present with sleep disturbances. The discovery that fatal familial insomnia is an inherited prion disease clearly widens the clinical spectrum of these disorders and raises the possibility that many other degenerative diseases of unknown etiology may be caused by prions (Johnson 1992; Medori *et al.* 1992*b*).

Other point mutations at codons 117, 198 and 217 also segregate with inherited prion diseases (Doh-ura *et al.* 1989; Hsiao *et al.* 1991*b*, 1992). Patients with a dementing or telencephalic form of GSS have a mutation at codon 117. These patients, as well as some in other families, were once thought to have familial Alzheimer's disease, but are now known to have prion diseases on the basis of PrP immunostaining of amyloid plaques and PrP gene mutations (Farlow *et al.* 1989; Ghetti *et al.* 1989; Giaccone *et al.* 1990; Nochlin *et al.* 1989). Patients with the codon 198 mutation have numerous neurofibrillary tangles that stain with antibodies to τ. They have amyloid plaques (Farlow *et al.* 1989; Ghetti *et al.* 1989; Giaccone *et al.* 1990; Nochlin *et al.* 1989) that are composed largely of a PrP fragment extending from residues 58 to 150 (Tagliavini *et al.* 1991). A genetic linkage study of this family produced a LOD score exceeding 6 (Dlouhy *et al.* 1992). The neuropathology of two patients of Swedish ancestry with the codon 217 mutation (Ikeda *et al.* 1991) was similar to that of patients with the codon 198 mutation.

At PrP codon 129, an amino acid (Met–Val) polymorphism (figure 3) has been identified (Owen *et al.* 1990). Patients with CJD after treatment with human pituitary growth hormone (Buchanan *et al.* 1991; Fradkin *et al.* 1991) or gonadotrophin have a significant preponderance of the Val allele (Collinge *et al.* 1991) compared with the general population. Sporadic CJD patients were found to be homozygous for the Met or Val allele at codon 129 but were rarely heterozygous (Palmer *et al.* 1991). The finding was interpreted (Hardy 1991; Palmer *et al.* 1991) as being consistent with the hypothesis for the existence of PrP^{c}/PrP^{Sc} heterodimers and that these forms feature in the replication of prions (Prusiner 1991; Prusiner *et al.* 1990; see § 10).

7. SPONTANEOUS NEURODEGENERATION IN TRANSGENIC MICE: ATTEMPTS TO DEMONSTRATE *DE NOVO* SYNTHESIS OF PRIONS

Transgenic modifications have been used to investigate the control of onset of infections, prion synthesis and neuropathology. When the codon 102 point mutation was introduced into MoPrP in transgenic (Tg) mice, spontaneous central nervous system (CNS) degeneration occurred, characterized by clinical signs indistinguishable from experimental murine scrapie and neuropathology consisting of widespread spongiform morphology and astrocytic gliosis (Hsiao *et al.* 1990). By inference, these results suggest that PrP mutations cause GSS and familial CJD. It is unclear whether low levels of protease-resistant PrP in the brains of Tg mice with the GSS mutation is PrP^{Sc}or residual PrP^{c}. Undetectable or low levels of PrP^{Sc} in the brains of these Tg mice are consistent with the results of transmission experiments that suggest low titres of infectious prions. Brain extracts transmit CNS degeneration to inoculated recipients, and the *de novo* synthesis of prions has been demonstrated by serial passage from one Tg (GSSMoPrP) mouse that developed spontaneous neurodegeneration (Hsiao *et al.* 1991*c*). If these observations can be supported by additional studies with similar results and the possibility of contamination eliminated, then it can be argued that prions are devoid of foreign nucleic acid, in accord with many studies that use other experimental approaches (Bellinger-Kawahara *et al.* 1987*a*,*b*; Diedrich *et al.* 1987; Diener *et al.* 1982; Duguid *et al.* 1988; Gabizon *et al.* 1988; Kellings *et al.* 1992; McKinley *et al.* 1983*b*; Meyer *et al.* 1991; Neary *et al.* 1991; Oesch *et al.* 1988).

One view of the PrP gene mutations has been that they render individuals susceptible to a common 'virus' (Aiken & Marsh 1990; Chesebro *et al.* 1985; Kimberlin 1990). In this scenario, the putative scrapie virus is thought to persist within a worldwide reservoir of humans, animals or insects without causing detectable illness. Yet 1 in 10^6 individuals develop sporadic CJD and die from a lethal 'infection' while $\sim 100\%$ of people with PrP point mutations or inserts appear eventually to develop neurologic dysfunction. That germline mutations found in the PrP genes of patients and at-risk individuals are the cause of familial prion diseases is supported by experiments with Tg(GSS MoPrP) mice described above (Hsiao & Prusiner 1990; Hsiao *et al.* 1991*c*; Weissmann 1991*b*). The Tg mouse studies also argue that sporadic CJD might arise from the spontaneous conversion of PrP^{C} to PrP^{CJD} due to either a somatic mutation of the PrP gene or a rare event involving modification of wild-type PrP^{C} (Prusiner 1991).

8. SPECIES BARRIERS IN THE TRANSMISSION OF PRION DISEASES

Passage of prions between species is a stochastic process characterized by prolonged incubation times (Pattison 1965, 1966; Pattison & Jones 1967). Prions synthesized *de novo* reflect the sequence of the host PrP gene and not that of the PrP^{Sc} molecules in the inoculum (Bockman *et al.* 1987). On subsequent passage in a homologous host, the incubation time shortens to that recorded for all subsequent passages and it becomes a non-stochastic process. The species barrier concept is of practical importance in assessing the risk for humans of developing CJD after consumption of scrapie-infected lamb or BSE beef.

To test the hypothesis that differences in PrP gene sequences might be responsible for the species barrier, Tg mice expressing SHaPrP were constructed (Prusiner *et al.* 1990; Scott *et al.* 1989). The PrP genes of Syrian hamsters and mice encode proteins differing at 16 positions. Incubation times in four lines of Tg(SHaPrP) mice inoculated with Mo prions were prolonged compared with those observed for non-Tg, control mice (figure 4*a*). Inoculation of Tg(SHaPrP) mice with SHa prions demonstrated abrogation of the species barrier resulting in abbreviated incubation times due to a non-stochastic process (figure 4*b*) (Prusiner *et al.* 1990; Scott *et al.* 1989). The length of the incubation time after inoculation with SHa prions was inversely proportional to the level of $SHaPrP^{C}$ in the brains of Tg(SHaPrP) mice (figure 4*b*,*c*) (Prusiner *et al.* 1990). $SHaPrP^{Sc}$ levels in the brains of clinically ill mice were similar in all four Tg(SHaPrP) lines inoculated with SHa prions (figure 4*d*). Bioassays of brain extracts from clinically ill Tg(SHaPrP) mice inoculated with Mo prions revealed that only Mo prions but no SHa prions were produced (figure 4*e*). Conversely, inoculation of Tg(SHaPrP) mice with SHa prions led to only the synthesis of SHa prions (figure 4*f*). Thus, the *de novo* synthesis of prions is species specific and reflects the genetic origin of the inoculated prions. Similarly, the neuropathology of Tg(SHaPrP) mice is determined by the genetic origin of prion inoculum. Mo prions injected into Tg(SHaPrP) mice produced a neuropathology characteristic of mice with scrapie. A moderate degree of vacuolation in both the grey and white matter was found but amyloid plaques were rarely detected (figure 4*g*). Inoculation of Tg(SHaPrP) mice with SHa prions produced intense vacuolation of the grey matter, sparing of the white matter, and numerous SHaPrP amyloid plaques characteristic of Syrian hamsters with scrapie (figure 4*h*).

9. PRION DIVERSITY

There is good evidence for multiple 'strains' or distinct isolates of prions as defined by specific incubation times, distribution of vacuolar lesions, and patterns of PrP^{Sc} accumulation (Bruce *et al.* 1989; Dickinson *et al.* 1968; Fraser & Dickinson, 1973; Hecker *et al.* 1992). The mechanism by which isolate-specific information is carried by prions remains enigmatic; indeed, explaining the molecular basis of prion diversity seems to be a formidable challenge. For many years, some investigators argued that scrapie is caused by a virus-like particle which contains a scrapie-specific nucleic acid that encodes the information expressed by each isolate (Bruce & Dickinson 1987). To date, no such polynucleotide has been identified by a wide variety of techniques including measurements of the nucleic acids in purified preparations. An alternative hypothesis has been suggested, where PrP^{Sc} alone is capable of transmitting disease but the characteristics of PrP^{Sc} might be modified by a cellular RNA (Weissman 1991*a*). This accessory cellular RNA is postulated to induce its own synthesis upon transmission from one host to another.

Two additional hypotheses not involving a nucleic acid have been offered to explain distinct prion isolates: a non-nucleic acid second component might create prion diversity, or post-translational modification of PrP^{Sc} might be responsible for the different properties of distinct prion isolates (Prusiner 1991). Whether the PrP^{Sc} modification is chemical or conformational alone remains to be established, but no candidate chemical modifications have been identified. Structural studies of the GPI anchors of two SHa isolates have failed to reveal any differences; interestingly, about 40% of the anchor glycans have sialic acid residues (Stal *et al.* 1992*a*). A portion of the PrP^{C} GPI anchors also has sialic acid residues; PrP is the first protein found to have sialic acid residues attached to GPI anchors.

Although the structures of Asn-linked carbohydrates have been analysed for PrP^{Sc} of one isolate (Endo *et al.* 1989), no data are available for PrP^{Sc} of other isolates or PrP^{C}. The great diversity of Asn-linked carbohydrates makes them candidates for isolate-specific information but there is no precedent for Asn-linked carbohydrates instructing the synthesis of more of the same compounds. In recent studies, we found that distinct isolates produce different, reproducible patterns of PrP^{Sc} accumulation (Hecker *et al.* 1992). These findings have given rise to the hypothesis that PrP^{Sc} synthesis occurs in particular sets of cells for a given distinct prion isolate. Whether different Asn-linked carbohydrates function to target PrP^{Sc} of a distinct isolate to a particular set of cells where the same Asn-linked carbohydrates will be coupled to PrP^{C} before its conversion to PrP^{Sc} remains to be established. Even though this hypothesis is attractive, it must be noted that PrP^{Sc} synthesis in scrapie-infected cells occurs in the presence of tunicamycin, which inhibits Asn-linked glycosylation, and with PrP molecules mutated at the Asn-linked glycosylation concensus sites (Taraboulos *et al.* 1990*a*). Whether SHa scrapie prions can be synthesized in Tg mice expressing SHaPrP with mutated Asn-linked glycosylation concensus sites and the properties exhibited by distinct isolates is currently under investigation. Of note, two different isolates from mink dying of trans-

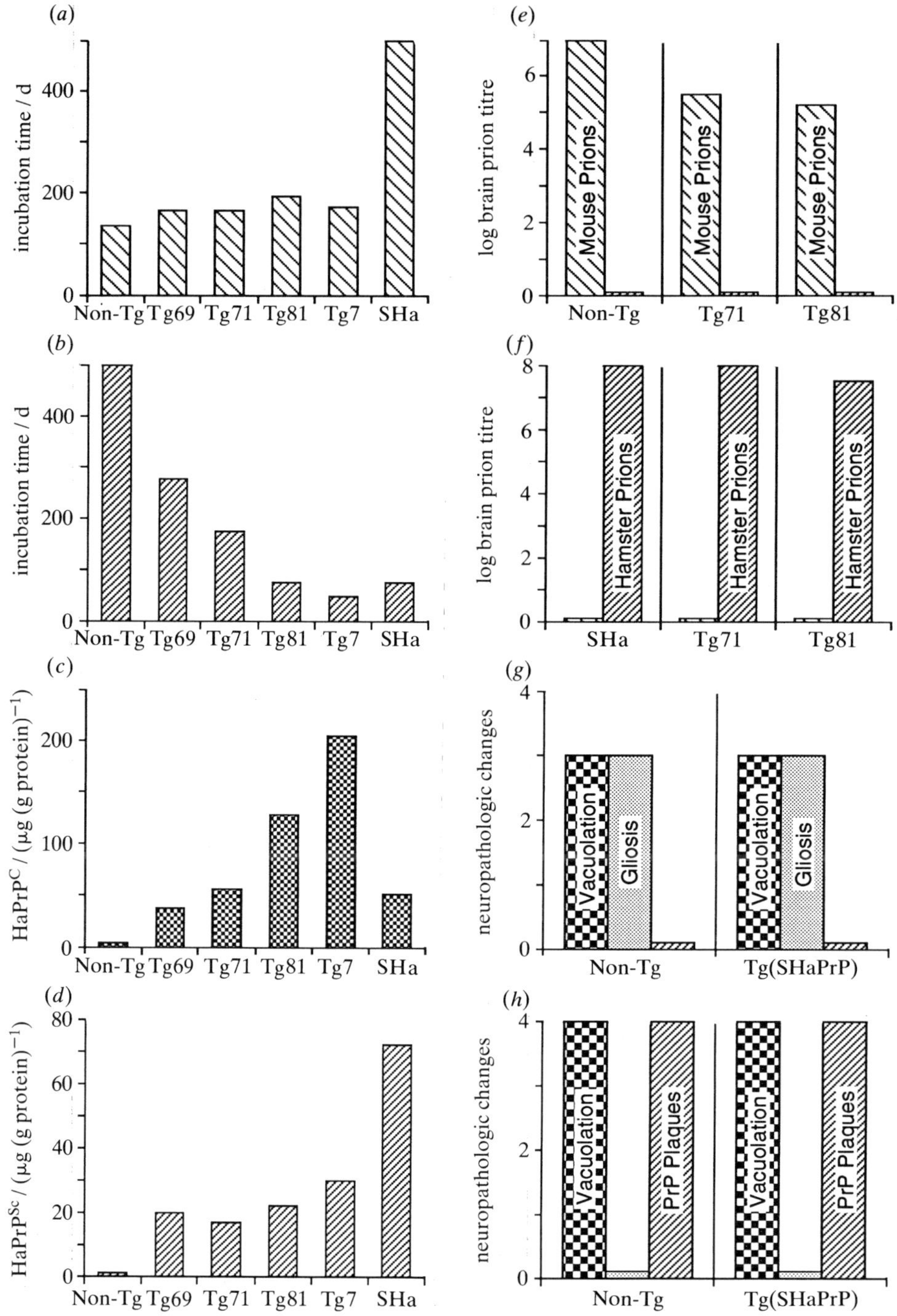

Figure 4. Transgenic mice expressing Syrian hamster (SHa) prion protein exhibit species-specific scrapie incubation times, infectious prion synthesis and neuropathology (Prusiner *et al.* 1990). (*a*) Scrapie incubation times in non-transgenic mice (Non-Tg) and four lines of transgenic mice expressing SHaPrP and Syrian hamsters inoculated intracerebrally with $\sim 10^6$ ID$_{50}$ units of Chandler Mo prions serially passaged in Swiss mice. The four lines of transgenic mice have different numbers of transgene copies: Tg69 and Tg71 mice have two to four copies of the SHaPrP transgene, whereas Tg81 have 30 to 50 copies and Tg7 mice have >60. Incubation times are number of days from inoculation to onset of neurologic dysfunction. (*b*) Scrapie incubation times in mice and hamsters inoculated with $\sim 10^7$ ID$_{50}$ units of Sc237 prions serially passaged in Syrian hamsters and as described in (*a*). (*c*) Brain SHaPrP^{C} in transgenic mice and hamsters. SHaPrP^{C} levels were quantitated by an enzyme-linked immunoassay. (*d*) Brain SHaPrP^{Sc} in transgenic mice and hamsters. Animals were killed after exhibiting clinical signs of scrapie. SHaPrP^{Sc} levels were determined by immunoassay. (*e*) Prion titres in brains of clinically ill animals after inoculation with Mo prions. Brain extracts from Non-Tg, Tg71, and Tg81 mice were bioassayed for prions in mice (left) and hamsters (right). (*d*) Prion titres in brains of clinically ill animals after inoculation with SHa prions. Brain extracts from Syrian hamsters as well as Tg71 and Tg81 mice were bioassayed for prions in mice (left) and hamsters (right). (*g*) Neuropathology in Non-Tg mice and Tg(SHaPrP) mice with clinical signs of scrapie after inoculation with Mo prions. Vacuolation in grey (left) and white (centre) matter; PrP amyloid plaques (right). Vacuolation score: 0 = none, 1 = rare, 2 = modest, 3 = moderate, 4 = intense. (*h*) Neuropathology in Syrian hamsters and transgenic mice inoculated with SHa prions. Degree of vacuolation and frequency of PrP amyloid plaques as described in (*g*). Adapted from Prusiner (*Science, Wash.* **252**, 1515–1522, 1991).

missible mink encephalopathy exhibit different sensitivities of PrP^{Sc} to proteolytic digestion, supporting the suggestion that isolate-specific information might be carried by PrP^{Sc} (Bessen & Marsh 1992*a*,*b*; Marsh *et al.* 1991).

10. PRION REPLICATION

Many experimental studies argue persuasively that prions are devoid of nucleic acid, yet the complete structure of the prion particle, as well as the mechanism by which prions multiply, remains to be established. Although the search for a scrapie-specific nucleic acid continues to be unrewarding, some investigators steadfastly cling to the notion that this putative polynucleotide drives prion replication. If prions are found to contain a scrapie-specific nucleic acid, then such a molecule would be expected to direct scrapie agent replication using a strategy similar to that used by viruses. In the absence of any chemical or physical evidence for a scrapie-specific polynucleotide (Aiken *et al.* 1990; Akowitz *et al.* 1990; Bellinger-Kawahara *et al.* 1987*a*,*b*; Diedrich *et al.* 1987; Diener *et al.* 1982; Duguid *et al.* 1988; Gabizon *et al.* 1988; Kellings *et al.* 1992; McKinley *et al.* 1983*b*; Meyer *et al.* 1991; Murdoch *et al.* 1990; Neary *et al.* 1991; Oesch *et al.* 1988), it seems reasonable to consider some alternative mechanisms that might feature in prion biosynthesis. The multiplication of prion infectivity is an exponential process in which the post-translational conversion of PrP^{C} or a precursor to PrP^{Sc} appears to be obligatory (Borchelt *et al.* 1990, 1992; Caughey & Raymond 1991).

Let us consider the remote possibility that prions do contain an as yet undetected polynucleotide, then, presumably, prion replication would involve a virus-like strategy. The putative scrapie-specific nucleic acid would act as a template for its own synthesis using cellular polymerases. By an as yet undefined mechanism, the putative scrapie-specific nucleic acid would stimulate the conversion of PrP^{C} to PrP^{Sc}. Although this putative scrapie-specific nucleic acid would provide a plausible explanation for prion diversity, it would require that this nucleotide sequence be able to discriminate between SHaPrP and MoPrP in Tg(SHaPrP) mice. In addition, the putative scrapie-specific nucleic acid would have to be ubiquitous to explain how sporadic CJD occurs with an incidence of 1 in 10^6 (Brown 1980; Masters *et al.* 1978) all over the planet whereas virtually all people carrying PrP gene mutations develop prion disease.

A more likely scenario is that prions do not contain a scrapie-specific nucleic acid; rather, they are composed entirely of PrP^{Sc} molecules. If this is the case, then the species barrier for prion transmission, the results with Tg(SHaPrP) mice, and infectious prions in the brains of patients with inherited prion diseases can be more readily explained. If prions are composed entirely of PrP^{Sc}, then replication must involve the interaction of nascent PrP^{C} or a precursor with PrP^{Sc} (Prusiner 1991; Prusiner *et al.* 1990). Although there are no physical data demonstrating the existence of PrP^{C}/PrP^{Sc} heterodimers, it is difficult to explain the results obtained with Tg(SHaPrP) mice in studies of prion replication. Moreover, other studies have shown that patients homologous for the Met–Val polymorphism at codon 129 are predisposed to sporadic CJD whereas those with heterozygous alleles at codon 129 are relatively protected (Palmer *et al.* 1991). These findings have been interpreted as being consistent with the hypothesis that prion replication is most efficient when the primary structures of PrP^{C} and PrP^{Sc} are the same. As noted above, although the PrP^{Sc} model is consistent with all of the experimental data, it continues to be problematic with respect to explaining the molecular basis of multiple distinct scrapie prion isolates or 'strains'.

The formal possibility remains that prions contain a second component which is not a nucleic acid. A small polypeptide, a polysaccharide, a lipid-glycan or a phospholipid–sterol complex are all possibilities, but there is no evidence for any of these molecules as prion components.

Some investigators have suggested that scrapie agent multiplication proceeds through a crystalization process involving PrP amyloid formation (Gajdusek 1988, 1990; Gajdusek & Gibbs 1990). Against this hypothesis is the absence or rarity of amyloid plaques in many prion diseases, as well as the inability to identify any amyloid-like polymers in cultured cells chronically synthesizing prions (McKinley *et al.* 1991*a*; Prusiner *et al.* 1990). Purified infectious preparations isolated from scrapie-infected hamster brains exist as amorphous aggregates; only if PrP^{Sc} is exposed to detergents and limited proteolysis, does it then polymerize into prion rods exhibiting the ultrastructural and tinctorial features of amyloid (McKinley *et al.* 1991*a*). Furthermore, dispersion of prion rods into detergent–lipid–protein complexes results in a 10- to 100-fold increase in scrapie titre and no rods could be identified in these fractions by electron microscopy (Gabizon *et al.* 1987).

11. CONCLUDING REMARKS

The study of prions has taken several unexpected directions over the past few years. The discovery that prion diseases in humans are uniquely both genetic and infectious has greatly strengthened and extended the prion concept. To date, 12 different mutations in the human PrP gene all resulting in non-conservative substitutions have been found to be either linked genetically to or segregate with the inherited prion diseases. Yet the transmissible prion particle is composed largely, if not entirely, of an abnormal isoform of the prion protein designated PrP^{Sc} (Prusiner 1991). These findings suggest that prion diseases should be considered pseudoinfections because the particles transmiting disease appear to be devoid of a foreign nucleic acid and thus differ from all known microorganisms as well as viruses and viroids. Because much information, especially about scrapie of rodents, has been derived using experimental protocols adapted from virology, we continue to use terms such as infection, incubation period, transmissibility and end-point titration in studies of prion diseases.

It seems likely that the principles learned from the study of prion diseases will be applicable to elucidating the causes of more common neurodegenerative diseases. Such disorders include Alzheimer's disease, amyotrophic lateral sclerosis and Parkinson's disease. Because people at risk for inherited prion diseases can now be identified decades before neurologic dysfunction is evident, the development of an effective therapy is imperative. If PrP^C can be diminished in humans without deleterious effects, as is the case for *Prn-p*$^{0/0}$ mice (Büeler *et al.* 1992), then reducing the level of PrP mRNA with antisence oligonucleotides might prove an effective therapeutic approach delaying the onset of CNS symptoms and signs.

The study of prion biology and diseases seems to be a new and emerging area of biomedical investigation. Although prion biology has its roots in virology, neurology and neuropathology, its relations to the disciplines of molecular and cell biology as well as protein chemistry have become evident only recently. It seems likely that learning how prions multiply and cause disease may open up new vistas into many areas of disease-related research.

Supported by grants from the National Institutes of Health (NS14069, AG08967, AG02132 and NS22786) and the American Health Assistance Foundation, as well as by gifts from Sherman Fairchild Foundation and National Medical Enterprises.

This manuscript is dedicated to Alan Williams, whose outstanding scientific contributions, warm friendship and exceptional intellect will be greatly missed. His untimely death saddens us all, and leaves a great void in the community of scientists and scholars, which he so enriched with his grace, zeal and enthusiasm.

REFERENCES

Aiken, J.M. & Marsh, R.F. 1990 The search for scrapie agent nucleic acid. *Microbiol. Rev.* **54**, 242–246.

Aiken, J.M., Williamson, J.L., Borchardt, L.M. & Marsh, R.F. 1990 Presence of mitochrondrial D-loop DNA in scrapie-infected brain preparations enriched for the prion protein. *J. Virol.* **64**, 3265–3268.

Akowitz, A., Sklaviadis, T., Manuelidis, E.E. & Manuelidis, L. 1990 Nuclease-resistant polyadenylated RNAs of significant size are detected by PCR in highly purified Creutzfeldt–Jakob disease preparations. *Microb. Pathog.* **9**, 33–45.

Alpers, M.P. 1979 Epidemiology and ecology of kuru. In *Slow transmissible diseases of the nervous system*, vol. 1 (ed. S. B. Prusiner & W. J. Hadlow), pp. 67–90. New York: Academic Press.

Alpers, M. 1987 Epidemiology and clinical aspects of kuru. In *Prions – novel infectious pathogens causing scrapie and Creutzfeldt-Jakob Disease* (ed. S. B. Prusiner & M. P. McKinley), pp. 451–465. Orlando: Academic Press.

Alter, M. & Kahana, E. 1976 Creutzfeldt–Jakob disease among Libyan Jews in Israel. *Science, Wash.* **192**, 428.

Baldwin, M.A., Stahl, N., Reinders, L.G., Gibson, B.W., Prusiner, S.B. & Burlingame, A.L. 1990 Permethylation and tandem mass spectrometry of oligosaccharides having free hexosamine: analysis of the glycoinositol phospholipid anchor glycan from the scrapie prion protein. *Analyt. Biochem.* **191**, 174–182.

Basler, K., Oesch, B., Scott, M., Westaway, D., Wälchli, M., Groth, D.F., McKinley, M.P., Prusiner, S.B. & Weissmann, C. 1986 Scrapie and cellular PrP isoforms are encoded by the same chromosomal gene. *Cell* **46**, 417–428.

Bazan, J.F., Fletterick, R.J., McKinley, M.P. & Prusiner, S.B. 1987 Predicted secondary structure and membrane topology of the scrapie prion protein. *Protein Eng.* **1**, 125–135.

Bellinger-Kawahara, C., Cleaver, J.E., Diener, T.O. & Prusiner, S.B. 1987*a* Purified scrapie prions resist inactivation by UV irradiation. *J. Virol.* **61**, 159–166.

Bellinger-Kawahara, C., Diener, T.O., McKinley, M.P., Groth, D.F., Smith, D.R. & Prusiner, S.B. 1987*b* Purified scrapie prions resist inactivation by procedures that hydrolyze, modify, or shear nucleic acids. *Virology* **160**, 271–274.

Bertoni, J.M., Brown, P., Goldfarb, L., Gajdusek, D. & Omaha, N.E. 1992 Familial Creutzfeldt–Jakob disease with the PRNP codon 200^{lys} mutation and supranuclear palsy but without myoclonus or periodic EEG complexes. *Neurology* **42**(4, Suppl. 3), 350. (Abstr.)

Bessen, R.A. & Marsh, R.F. 1992*a* Biochemical and physical properties of the prion protein from two strains of the transmissible mink encephalopathy agent. *J. Virol.* **66**, 2096–2101.

Bessen, R.A. & Marsh, R.F. 1992*b* Identification of two biologically distinct strains of transmissible mink encephalopathy in hamsters. *J. gen. Virol.* **73**, 329–334.

Blum, B., Bakalara, N. & Simpson, L. 1990 A model for RNA editing in kinetoplastid mitochondria: "guide" RNA molecules transcribed from maxicircle DNA provide edited information. *Cell* **60**, 189–198.

Bockman, J.M., Prusiner, S.B., Tateishi, J. & Kingsbury, D.T. 1987 Immunoblotting of Creutzfeldt–Jakob disease prion proteins: host species-specific epitopes. *Ann. Neurol.* **21**, 589–595.

Bolton, D.C., McKinley, M.P. & Prusiner, S.B. 1982 Identification of a protein that purifies with the scrapie prion. *Science, Wash.* **218**, 1309–1311.

Bolton, D.C., Meyer, R.K. & Prusiner, S.B. 1985 Scrapie PrP 27-30 is a sialoglycoprotein. *J. Virol.* **53**, 596–606.

Borchelt, D.R., Scott, M., Taraboulos, A., Stahl, N. & Prusiner, S.B. 1990 Scrapie and cellular prion proteins differ in their kinetics of synthesis and topology in cultured cells. *J. Cell Biol.* **110**, 743–752.

Borchelt, D.R., Taraboulos, A. & Prusiner, S.B. 1992 Evidence for synthesis of scrapie prion proteins in the endocytic pathway. *J. biol. Chem.* **267**, 6188–6199.

Brown, P. 1980 An epidemiologic critique of Creutfeldt–Jakob disease. *Epidemiol. Rev.* **2**, 113–135.

Brown, P., Goldfarb, L.G. & Gajdusek, D.C. 1991 The new biology of spongiform encephalopathy: infectious amyloidoses with a genetic twist. *Lancet* **337**, 1019–1022.

Brown, P., Goldfarb, L.G., Kovanen, J., Haltia, M., Cathala, F., Sulima, M., Gibbs, C.J.J. & Gajdusek, D.C. 1992 Phenotypic characteristics of familial Creutzfeldt–Jakob disease associated with the codon 178^{Asn} *PRNP* mutation. *Ann. Neurol.* **31**, 282–285.

Bruce, M.E. & Dickinson, A.G. 1987 Biological evidence that the scrapie agent has an independent genome. *J. gen. Virol.* **68**, 79–89.

Bruce, M.E., McBride, P.A. & Farquhar, C.F. 1989 Precise targeting of the pathology of the sialoglycoprotein, PrP, and vacuolar degeneration in mouse scrapie. *Neurosci. Lett.* **102**, 1–6.

Buchanan, C.R., Preece, M.A. & Milner, R.D.G. 1991 Mortality, neoplasia and Creutzfeldt–Jakob disease in patients treated with pituitary growth hormone in the United Kingdom. *Br. med. J.* **302**, 824–828.

Büeler, H., Fischer, M., Lang, Y., Blüthmann, H., Lipp, H.-L., DeArmond, S.J., Prusiner, S.B., Aguet, M. & Weissmann, C. 1992 The neuronal cell surface protein PrP is not essential for normal development and behaviour of the mouse. *Nature, Lond.* **356**, 577–582.

Butler, D.A., Scott, M.R.D., Bockman, J.M., Borchelt, D.R., Taraboulos, A., Hsiao, K.K., Kingsbury, D.T. & Prusiner, S.B. 1988 Scrapie-infected murine neuroblastoma cells produce protease-resistant prion proteins. *J. Virol.* **62**, 1558–1564.

Carlson, G.A., Kingsbury, D.T., Goodman, P.A., Coleman, S., Marshall, S.T., DeArmond, S.J., Westaway, D. & Prusiner, S.B. 1986 Linkage of prion protein and scrapie incubation time genes. *Cell* **46**, 503–511.

Carlson, G.A., Goodman, P.A., Lovett, M., Taylor, B.A., Marshall, S.T., Peterson-Torchia, M., Westaway, D. & Prusiner, S.B. 1988 Genetics and polymorphism of the mouse prion gene complex: the control of scrapie incubation time. *Molec. Cell Biol.* **8**, 5528–5540.

Caughey, B. & Raymond, G.J. 1991 The scrapie-associated form of PrP is made from a cell-surface precursor that is both protease- and phospholipase-sensitive. *J. biol. Chem.* **266**, 18217–18223.

Caughey, B., Race, R.E., Ernst, D., Buchmeier, M.J. & Chesebro, B. 1989 Prion protein biosynthesis in scrapie-infected and uninfected neuroblastoma cells. *J. Virol.* **63**, 175–181.

Caughey, B., Raymond, G.J., Ernst, D. & Race, R.E. 1991 N-terminal truncation of the scrapie-associated form of PrP by lysosomal protease(s): implications regarding the site of conversion of PrP to the protease-resistant state. *J. Virol.* **65**, 6597–6603.

Chesebro, B., Race, R., Wehrly, K., Nishio, J., Bloom, M., Lechner, D., Bergstrom, S., Robbins, K., Mayer, L., Keith, J.M., Garon, C. & Haase, A. 1985 Identification of scrapie prion protein-specific mRNA in scrapie-infected and uninfected brain. *Nature, Lond.* **315**, 331–333.

Collinge, J., Harding, A.E., Owen, F., Poulter, M., Lofthouse, R., Boughey, A.M., Shah, T. & Crow, T.J. 1989 Diagnosis of Gerstmann–Sträussler syndrome in familial dementia with prion protein gene analysis. *Lancet* (ii), 15–17.

Collinge, J., Owen, F., Poulter, H., Leach, M., Crow, T., Rosser, M., Hardy, J., Mullan, H., Janota, I. & Lantos, P. 1990 Prion dementia without characteristic pathology. *Lancet* **336**, 7–9.

Collinge, J., Palmer, M.S. & Dryden, A.J. 1991 Genetic predisposition to iatrogenic Creutzfeldt–Jakob disease. *Lancet* **337**, 1441–1442.

Creutzfeldt, H.G. 1920 Über eine eigenartige herdförmige Erkrankung des Zentralnervensystems. *Z. ges. Neurol. Psychiat.* **57**, 1–18.

Crow, T.J., Collinge, J., Ridley, R.M., Baker, H.F., Lofthouse, R., Owen, F. & Harding, A.E. 1990 Mutations in the prion gene in human transmissible dementia. In *Seminar on Molecular Approaches to Research in Spongiform Encephalopathies in Man.* London: Medical Research Council. (Abstr.)

Cuillé, J. & Chelle, P.L. 1939 Experimental transmission of trembling to the goat. *C.r. hebd. Séanc. Acad. Sci., Paris* **208**, 1058–1060.

Dealler, S.F. & Lacey, R.W. 1990 Transmissible spongiform encephalopathies: the threat of BSE to man. *Food Microbiol.* **7**, 253–279.

DeArmond, S.J., McKinley, M.P., Barry, R.A., Braunfeld, M.B., McColloch, J.R. & Prusiner, S.B. 1985 Identification of prion amyloid filaments in scraple-infected brain. *Cell* **41**, 221–235.

DeArmond, S.J., Mobley, W.C., DeMott, D.L., Barry, R.A., Beckstead, J.H. & Prusiner, S.B. 1987 Changes in the localization of brain prion proteins during scrapie infection. *Neurology* **37**, 1271–1280.

Dickinson, A.G., Meikle, V.M.H. & Fraser, H. 1968 Identification of a gene which controls the incubation period of some strains of scrapie agent in mice. *J. comp. Pathol.* **78**, 293–299.

Dickinson, A.G., Young, G.B., Stamp, J.T. & Renwick, C.C. 1965 An analysis of natural scrapie in Suffolk sheep. *Heredity* **20**, 485–503.

Diedrich, J., Weitgrefe, S., Zupancic, M., Staskus, K., Retzel, E., Haase, A.T. & Race, R. 1987 The molecular pathogenesis of astrogliosis in scrapie and Alzheimer's disease. *Microb. Pathog.* **2**, 435–442.

Diener, T.O., McKinley, M.P. & Prusiner, S.B. 1982 Viroids and prions. *Proc. natn. Acad. Sci. U.S.A.* **79**, 5220–5224.

Dlouhy, S.R., Hsiao, K., Farlow, M.R., Foroud, T., Conneally, P.M., Johnson, P., Prusiner, S.B., Hodes, M.E. & Ghetti, B. 1992 Linkage of the Indiana kindred of Gerstmann–Straussler–Scheinker disease to the prion protein gene. *Nature Genet.* **1**, 64–67.

Doh-ura, K., Tateishi, J., Sasaki, H., Kitamoto, T. & Sakaki, Y. 1989 Pro→Leu change at position 102 of prion protein is the most common but not the sole mutation related to Gerstmann–Sträussler syndrome. *Biochem. biophys. Res. Commun.* **163**, 974–979.

Doms, R.W., Russ, G. & Yewdell, J.W. 1989 Brefeldin A redistributes resident and itinerant Golgi proteins to the endoplasmic reticulum. *J. Cell Biol.* **109**, 61–72.

Donehower, L.A., Harvey, M., Slagle, B.L., McArthur, M.J., Montgomery, C.A.J., Butel, J.A. & Bradley, A. 1992 Mice deficient for p53 are developmentally normal but susceptible to spontaneous tumours. *Nature, Lond.* **356**, 215–221.

Duguid, J.R., Rohwer, R.G. & Seed, B. 1988 Isolation of cDNAs of scrapie-modulated RNAs by subtractive hybridization of a cDNA library. *Proc. natn. Acad. Sci. U.S.A.* **85**, 5738–5742.

Endo, T., Groth, D., Prusiner, S.B. & Kobata, A. 1989 Diversity of oligosaccharide structures linked to asparagines of the scrapie prion protein. *Biochemistry* **28**, 8380–8388.

Farlow, M.R., Yee, R.D., Dlouhy, S.R., Conneally, P.M., Azzarelli, B. & Ghetti, B. 1989 Gerstmann–Sträussler–Scheinker disease. I. Extending the clinical spectrum. *Neurology* **39**, 1446–1452.

Fink, J.K., Warren, J.T. Jr, Drury, I., Murman, D. & Peacock, B.A. 1991 Allele-specific sequencing confirms novel prion gene polymorphism in Creutzfeldt–Jakob disease. *Neurology* **41**, 1647–1650.

Fradkin, J.E., Schonberger, L.B., Mills, J.L., Gunn, W.J., Piper, J.M., Wysowski, D.K., Thomson, R., Durako, S. & Brown, P. 1991 Creutzfeldt–Jakob disease in pituitary growth hormone recipients in the United States. *J. Am. med. Ass.* **265**, 880–884.

Fraser, H. & Dickinson, A.G. 1973 Scrapie in mice. Agent-strain differences in the distribution and intensity of grey matter vacuolation. *J. comp. Pathol.* **83**, 29–40.

Fraser, H. & Dickinson, A.G. 1985 Targeting of scrapie lesions and spread of agent via the retino-tectal projection. *Brain Res.* **346**, 32–41.

Gabizon, R. & Prusiner, S.B. 1990 Prion liposomes. *Biochem. J.* **266**, 1–14.

Gabizon, R., McKinley, M.P. & Prusiner, S.B. 1987 Purified prion proteins and scrapie infectivity copartition into liposomes. *Proc. natn. Acad. Sci. U.S.A.* **84**, 4017–4021.

Gabizon, R., McKinley, M.P., Groth, D.F., Kenaga, L. & Prusiner, S.B. 1988 Properties of scrapie prion liposomes. *J. biol. Chem.* **263**, 4950–4955.

Gabizon, R., Meiner, Z., Cass, C., Kahana, E., Kahana, I.,

Avrahami, D., Abramsky, O., Scarlato, G., Prusiner, S.B. & Hsiao, K.K. 1991 Prion protein gene mutation in Libyan Jews with Creutzfeldt–Jakob disease. *Neurology* **41**, 160. (Abstr.)

Gabriel, J.-M., Oesch, B., Kretzschmar, H., Scott, M. & Prusiner, S.B. 1992 Molecular cloning and structural analysis of a candidate chicken prion protein. *Proc. natn. Acad. Sci. U.S.A.* **89**, 9097–9101.

Gajdusek, D.C. 1977 Unconventional viruses and the origin and disappearance of kuru. *Science, Wash.* **197**, 943–960.

Gajdusek, D.C. 1988 Transmissible and non-transmissible amyloidoses: autocatalytic post-translational conversion of host precursor proteins to β-pleated sheet configurations. *J. Neuroimmunol.* **20**, 95–110.

Gajdusek, D.C. 1990 Subacute spongiform encephalopathies: transmissible cerebral amyloidoses caused by unconventional viruses. In *Virology*, 2nd edn (ed. B. N. Fields, D. M. Knipe, R. M. Chanock, M. S. Hirsch, J. L. Melnick, T. P. Monath & B. Roizman), pp. 2289–2324. New York: Raven Press.

Gajdusek, D.C. & Gibbs, C.J. Jr 1990 Brain amyloidoses-precursor proteins and the amyloids of transmissible and nontransmissible dementias: scrapie–kuru–CJD viruses as infectious polypeptides or amyloid enhancing vactor. In *Biomedical advances in aging* (ed. A. Goldstein), pp. 3–24. New York: Plenum Press.

Gajdusek, D.C., Gibbs, C.J. Jr. & Alpers, M. 1966 Experimental transmission of a kuru-like syndrome to chimpanzees. *Nature, Lond.* **209**, 794–796.

Ghetti, B., Tagliavini, F., Masters, C.L., Beyreuther, K., Giaccone, G., Verga, L., Farlo, M.R., Conneally, P.M., Dlouhy, S.R., Azzarelli, B. & Bugiani, O. 1989 Gerstmann-Sträussler-Scheinker disease. II. Neurofibrillary tangles and plaques with PrP-amyloid coexist in an affected family. *Neurology* **39**, 1453–1461.

Giaccone, G., Tagliavini, F., Verga, L., Frangione, B., Farlow, M.R., Bugiani, O. & Ghetti, B. 1990 Neurofibrillary tangles of the Indiana kindred of Gerstmann–Sträussler–Scheinker disease share antigenic determinants with those of Alzheimer disease. *Brain Res.* **530**, 325–329.

Gibbons, R.A. & Hunter, G.D. 1967 Nature of the scrapie agent. *Nature, Lond.* **215**, 1041–1043.

Gibbs, C.J. Jr, Gajdusek, D.C., Asher, D.M., Alpers, M.P., Beck, E., Daniel, P.M. & Matthews, W.B. 1968 Creutzfeldt–Jakob disease (spongiform encephalopathy): transmission to the chimpanzee. *Science, Wash.* **161**, 388–389.

Gibbs, C.J. Jr, Joy, A., Heffner, R., Franko, M., Miyazaki, M., Asher, D.M., Parisi, J.E., Brown, P.W. & Gajdusek, D.C. 1985 Clinical and pathological features and laboratory confirmation of Creutzfelt–Jakob disease in a recipient of pituitary-derived human growth hormone. *N. Engl. J. Med.* **313**, 734–738.

Goldfarb, L., Brown, P., Goldgaber, D., Garruto, R., Yanaghiara, R., Asher, D. & Gajdusek, D.C. 1990*a* Identical mutation in unrelated patients with Creutzfeldt–Jakob disease. *Lancet* **336**, 174–175.

Goldfarb, L., Korczyn, A., Brown, P., Chapman, J. & Gajdusek, D.C. 1990*b* Mutation in codon 200 of scrapie amyloid precursor gene linked to Creutzfeldt–Jakob disease in Sephardic Jews of Libyan and non-Libyan origin. *Lancet* **336**, 637–638.

Goldfarb, L.G., Brown, P., Goldgaber, D., Asher, D.M., Rubenstein, R., Brown, W.T., Piccardo, P., Kascsak, R.J., Boellaard, J.W. & Gajdusek, D.C. 1990*c* Creutzfeldt–Jakob disease and kuru patients lack a mutation consistently found in the Gerstmann–Sträussler–Scheinker syndrome. *Expl Neurol.* **108**, 247–250.

Goldfarb, L.G., Mitrova, E., Brown, P., Toh, B.H. & Gajdusek, D.C. 1990*d* Mutation in codon 200 of scrapie amyloid protein gene in two clusters of Creutzfeldt–Jakob disease in Slovakia. *Lancet* **336**, 514–515.

Goldfarb, L.G., Brown, P., McCombie, W.R., Goldgaber, D., Swergold, G.D., Wills, P.R., Cervenakova, L., Baron, H., Gibbs, C.J.J. & Gajdusek, D.C. 1991*a* Transmissible familial Creutzfeldt–Jakob disease associated with five, seven, and eight extra octapeptide coding repeats in the *PRNP* gene. *Proc. natn. Acad. Sci. U.S.A.* **88**, 10926–10930.

Goldfarb, L.G., Brown, P., Mitrova, E., Cervenakova, L., Goldin, L., Korczyn, A.D., Chapman, J., Galvez, S., Cartier, L., Rubenstein, R. & Gajdusek, D.C. 1991*b* Creutzfeldt–Jacob disease associated with the PRNP codon 200^{Lys} mutation: an analysis of 45 families. *Eur. J. Epidemiol.* **7**, 477–486.

Goldfarb, L.G., Haltia, M., Brown, P., Nieto, A., Kovanen, J., McCombie, W.R., Trapp, S. & Gajdusek, D.C. 1991*c* New mutation in scrapie amyloid precursor gene (at codon 178) in Finnish Creutzfeldt–Jakob kindred. *Lancet* **337**, 425.

Goldfarb, L.G., Brown, P., Haltia, M., Cathala, F., McCombie, W.R., Kovanen, J., Cervenakova, L., Goldin, L., Nieto, A., Godec, M.S., Asher, D.M. & Gajdusek, D.C. 1992 Creutzfeldt–Jakob disease cosegregates with the codon 178^{Asn} *PRNP* mutation in families of European origin. *Ann. Neurol.* **31**, 274–281.

Goldgaber, D., Goldfarb, L.G., Brown, P., Asher, D.M., Brown, W.T., Lin, S., Teener, J.W., Feinstone, S.M., Rubenstein, R., Kascsak, R.J., Boellaard, J.W. & Gajdusek, D.C. 1989 Mutations in familial Creutzfeldt–Jakob disease and Gerstmann–Sträussler–Scheinker's syndrome. *Expl Neurol.* **106**, 204–206.

Gordon, W.S. 1946 Advances in veterinary research. *Vet. Res.* **58**, 516–520.

Gordon, W.S. 1966 Variation in susceptibility of sheep to scrapie and genetic implications. In *Report of scrapie Seminar, ARS 91-53*, pp. 53–67. Washington, D.C.: U.S. Department of Agriculture.

Griffith, J.S. 1967 Self-replication and scrapie. *Nature, Lond.* **215**, 1043–1044.

Hadlow, W.J. 1959 Scrapie and kuru. *Lancet* (ii), 289–290.

Haltia, M., Kovanen, J., Goldfarb, L.G., Brown, P. & Gajdusek, D.C. 1991 Familial Creutzfeldt–Jakob disease in Finland: Epidemiological, clinical, pathological and molecular genetic studies. *Eur. J. Epidemiol.* **7**, 494–500.

Haraguchi, T., Fisher, S., Olofsson, S., Endo, T., Groth, D., Tarantino, A., Borchelt, D.R., Teplow, D., Hood, L., Burlingame, A., Lycke, E., Kobata, A. & Prusiner, S.B. 1989 Asparagine-linked glycosylation of the scrapie and cellular prion proteins. *Archs Biochem. Biophys.* **274**, 1–13.

Hardy, J. 1991 Prion dimers – a deadly duo. *Trends Neurosci.* **14**, 423–424.

Harris, D.A., Falls, D.L., Johnson, F.A. & Fischbach, G.D. 1991 A prion-like protein from chicken brain copurifies with an acetylcholine receptor-inducing activity. *Proc. natn. Acad. Sci. U.S.A.* **88**, 7664–7668.

Hay, B., Barry, R.A., Lieberburg, I., Prusiner, S.B. & Lingappa, V.R. 1987*a* Biogenesis and transmembrane orientation of the cellular isoform of the scrapie prion protein. *Molec. Cell Biol.* **7**, 914–920.

Hay, B., Prusiner, S.B. & Lingappa, V.R. 1987*b* Evidence for a secretory form of the cellular prion protein. *Biochemistry* **26**, 8110–8115.

Hecker, R., Taraboulos, A., Scott, M., Pan, K.-M., Torchia, M., Jendroska, K., DeArmond, S.J. & Prusiner, S.B. 1992 Replication of distinct prion isolates is region specific in brains of transgenic mice and hamsters. *Genes & Development.* **6**, 1213–1228.

Herzberg, L., Herzberg, B.N., Gibbs, C.J. Jr, Sullivan, W.,

Amyx, H. & Gajdusek, D.C. 1974 Creutzfeldt–Jakob disease: hypothesis for high incidence in Libyan Jews in Israel. *Science, Wash.* **186**, 848.

Hsiao, K. & Prusiner, S.B. 1990 Inherited human prion diseases. *Neurology* **40**, 1820–1827.

Hsiao, K., Baker, H.F., Crow, T.J., Poulter, M., Owen, F., Terwilliger, J.D., Westaway, D., Ott, J. & Prusiner, S.B. 1989*a* Linkage of a prion protein missense variant to Gerstmann–Sträussler syndrome. *Nature, Lond.* **338**, 342–345.

Hsiao, K.K., Doh-ura, K., Kitamoto, T., Tateishi, J. & Prusiner, S.B. 1989*b* A prion protein amino acid substitution in ataxic Gerstmann–Sträussler syndrome. *Ann. Neurol.* **26**, 137.

Hsiao, K.K., Scott, M., Foster, D., Growth, D.F., DeArmond, S.J. & Prusiner, S.B. 1990 Spontaneous neurodegeneration in transgenic mice with mutant prion protein of Gerstmann–Sträussler syndrome. *Science, Wash.* **250**, 1587–1590.

Hsiao, K., Meiner, Z., Kahana, E., Cass, C., Kahana, I., Avrahami, D., Scarlato, G., Abramsky, O., Prusiner, S.B. & Gabizon, R. 1991*a* Mutation of the prion protein in Libyan Jews with Creutzfeldt–Jakob disease. *N. Engl. J. Med.* **324**, 1091–1097.

Hsiao, K.K., Cass, C., Schellenberg, G.D., Bird, T., Devine-Gage, E. & Prusiner, S.B. 1991*b* A prion protein variant in a family with the telencephalic form of Gerstmann–Sträussler–Scheinker syndrome. *Neurology* **41**, 681–684.

Hsiao, K.K., Groth, D., Scott, M., Yang, S.-L., Serban, A., Rapp, D., Foster, D., Torchia, M., DeArmond, S.J. & Prusiner, S.B. 1991*c* Neurologic disease of transgenic mice which express GSS mutant prion protein is transmissible to inoculated recipient animals. In *Prion diseases of humans and animals Symposium*, London, Sept. 2–4, 1991. (Abstr.)

Hsiao, K., Dloughy, S., Ghetti, B., Farlow, M., Cass, C., Da Costa, M., Conneally, M., Hodes, M.E. & Prusiner, S.B. 1992 Mutant prion proteins in Gerstmann–Straussler–Scheinker disease with neurofibrillary tangles. *Nature Genet.* **1**, 68–71.

Hunter, N., Hope, J., McConnell, I. & Dickinson, A.G. 1987 Linkage of the scrapie-associated fibril protein (PrP) gene and Sinc using congenic mice and restriction fragment length polymorphism analysis. *J. gen. Virol.* **68**, 2711–2716.

Hunter, N., Foster, J.D., Dickinson, A.G. & Hope, J. 1989 Linkage of the gene for the scrapie-associated fibril protein (PrP) to the Sip gene in Cheviot sheep. *Vet. Rec.* **124**, 364–366.

Ikeda, S., Yanagisawa, N., Allsop, D. & Glenner, G.G. 1991 A variant of Gerstmann-Sträussler-Scheinker disease with β-protein epitopes and dystrophic neurites in the peripheral regions of PrP-immunoreactive amyloid plaques. In *Amyloid and amyloidosis 1990* (ed. J. B. Natvig, O. Forre, G. Husby, A. Husebekk, B. Skogen, K. Sletten & P. Westermark), pp. 737–740. Dordrecht: Kluwer Academic Publishers.

Jakob, A. 1921*a* Über eigenartige Erkrankungen des Zentralnervensystems mit bemerkenswertem anatomischen Befunde (spastische Pseudosclerose-Encephalomyelopathie mit disseminierten Degenerationsherden). Preliminary communication. *Dt. Z. Nervenheilk.* **70**, 132–146.

Jakob, A. 1921*b* Über eigenartige Erkrankungen des Zentralnervensystems mit bemerkenswertem anatomischen Befunde (spastische Pseudosklerose-Encephalomyelopathie mit disseminierten Degenerationsherden), *Z. ges. Neurol. Psychiat.* **64**, 147–228.

Jakob, A. 1921*c* Über eine der multiplen Sklerose klinisch nahestehende Erkrankung des Zentralnervensystems (spastische Pseudosklerose) mit bemerkenswertem anatomischem Befunde. Mitteilung eines vierten Falles. *Med. Klin.* **17**, 372–376.

Jendroska, K., Heinzel, F.P., Torchia, M., Stowring, L., Kretzshmar, H.A., Kon, A., Stern, A., Prusiner, S.B. & DeArmond, S.J. 1991 Proteinase-resistant prion protein accumulation in Syrian hamster brain correlates with regional pathology and scrapie infectivity. *Neurology* **41**, 1482–1490.

Johnson, R.T. 1992 Prion disease. *N. Engl. J. Med.* **326**, 486–487.

Kahana, E., Milton, A., Braham, J. & Sofer, D. 1974 Creutzfeldt–Jakob disease: focus among Libyan Jews in Israel. *Science, Wash.* **183**, 90–91.

Kane, P.M., Yamashiro, C.T., Wolczyk, D.F., Neff, N., Goebl, M. & Stevens, T.H. 1990 Protein splicing converts the yeast *TFP1* gene product to the 69-kD subunit of the vacuolar H^+-adenosine triphosphatase. *Science, Wash.* **250**, 651–657.

Kellings, K., Meyer, N., Mirenda, C., Prusiner, S.B. & Riesner, D. 1992 Further analysis of nucleic acids in purified scrapie prion preparations by improved return refocussing gel electrophoresis (RRGE). *J. Virol.* **73**, 1025–1029.

Kimberlin, R.H. 1990 Scrapie and possible relationships with viroids. *Semin. Virol.* **1**, 153–162.

Kimberlin, R.H., Field, H.J. & Walker, C.A. 1983 Pathogenesis of mouse scrapie: evidence for spread of infection from central to peripheral nervous system. *J. gen. Virol.* **64**, 713–716.

Kitamoto, T., Tateishi, J., Tashima, I., Takeshita, I., Barry, R.A., DeArmond, S.J. & Prusiner, S.B. 1986 Amyloid plaques in Creutzfeldt–Jakob disease stain with prion protein antibodies. *Ann. Neurol.* **20**, 204–208.

Klatzo, I., Gajdusek, D.C. & Zigas, V. 1959 Pathology of kuru. *Lab. Invest.* **8**, 799–847.

Kretzschmar, H.A., Prusiner, S.B., Stowring, L.E. & De Armond, S.J. 1986*a* Scrapie prion proteins are synthesized in neurons. *Am. J. Path.* **122**, 1–5.

Kretzschmar, H.A., Stowring, L.E., Westaway, D., Stubblebine, W.H., Prusiner, S.B. & DeArmond, S.J. 1986*b* Molecular cloning of a human prion protein cDNA. *DNA* **5**, 315–324.

Kretzschmar, H.A., Honold, G., Seitelberger, F., Feucht, M., Wessely, P., Mehraein, P. & Budka, H. 1991*a* Prion protein mutation in family first reported by Gerstmann, Straussler, and Scheinker. *Lancet* **337**, 1160. (Letter.)

Kretzschmar, H.A., Kufer, P., Riethmuller, G., DeArmond, S.J., Prusiner, S.B. & Schiffer, D. 1991*b* Prion protein mutation at codon 102 in an Italian family with Gerstmann–Sträussler–Scheinker syndrome. *Neurology* **42**, 809–810.

Laplanche, J.-L., Chatelain, J., Launay, J.-M., Gazengel, C. & Vidaud, M. 1990 Deletion in prion protein gene in a Moroccan family. *Nucl. Acids. Res.* **18**, 6745.

Lippincott-Schwartz, J., Yuan, L.C., Bonifacino, J.S. & Klausner, R.D. 1989 Rapid redistribution of Golgi proteins into the ER in cells treated with Brefeldin A: evidence for membrane cycling from the Golgi to ER. *Cell* **56**, 801–813.

Lopez, C.D., Yost, C.S., Prusiner, S.B., Myers, R.M. & Lingappa, V.R. 1990 Unusual topogenic sequence directs prion protein biogenesis. *Science, Wash.* **248**, 226–229.

M'Gowan, J.P. 1914 *Investigation into the disease of sheep called "scrapie"*. (114 pages.) Edinburgh: William Blackwood and Sons.

Manuelidis, L., Valley, S. & Manuelidis, E.E. 1985 Specific proteins associated with Creutzfeldt-Jakob disease and scrapie share antigenic and carbohydrate determinants. *Proc. natn. Acad. Sci. U.S.A.* **82**, 4263–4267.

Marsh, R.F., Bessen, R.A., Lehmann, S. & Hartsough, G.R. 1991 Epidemiological and experimental studies on a new incident of transmissible mink encephalopathy. *J. gen. Virol.* **72**, 589–594.

Masters, C.L., Harris, J.O., Gajdusek, D.C., Gibbs, C.J. Jr, Bernouilli, C. & Asher, D.M. 1978 Creutzfeldt–Jakob disease: patterns of worldwide occurrence and the significance of familial and sporadic clustering. *Ann. Neurol.* **5**, 177–188.

Masters, C.L., Gajdusek, D.C., Gibbs, C.J. Jr, Bernouilli, C. & Asher, D.M. 1979 Familial Creutzfeldt-Jakob disease and other familial dementias: an inquiry into possible models of virus-induced familial diseases. In *Slow transmissible diseases of the nervous system*, vol. 1 (ed. S. B. Prusiner & W. J. Hadlow), pp. 143–194. New York: Academic Press.

Masters, C.L., Gajdusek, D.C. & Gibbs, C.J. Jr 1981*a* Creutzfeldt–Jakob disease virus isolations from the Gerstmann–Sträussler syndrome. *Brain* **104**, 559–588.

Masters, C.L., Gajdusek, D.C. & Gibbs, C.J. Jr 1981*b* The familial occurrence of Creutzfeldt–Jakob disease and Alzheimer's disease. *Brain* **104**, 535–558.

McKinley, M.P., Bolton, D.C. & Prusiner, S.B. 1983*a* A protease-resistant protein is a structural component of the scrapie prion. *Cell* **35**, 57–62.

McKinley, M.P., Masiarz, F.R., Isaacs, S.T., Hearst, J.E. & Prusiner, S.B. 1983*b* Resistance of the scrapie agent to inactivation by psoralens. *Photochem. Photobiol.* **37**, 539–545.

McKinley, M.P., Meyers, R., Kenaga, L., Rahbar, F., Cotter, R., Serban, A. & Prusiner, S.B. 1991*a* Scrapie prion rod formation *in vitro* requires both detergent extraction and limited proteolysis. *J. Virol.* **65**, 1440–1449.

McKinley, M.P., Taraboulos, A., Kenaga, L., Serban, D., Stieber, A., DeArmond, S.J., Prusiner, S.B. & Gonatas, N. 1991*b* Ultrastructural localization of scrapie prion proteins in cytoplasmic vesicles of infected cultured cells. *Lab. Invest.* **65**, 622–630.

McKnight, S. & Tjian, R. 1986 Transcriptional selectivity of viral genes in mammalian cells. *Cell* **46**, 795–805.

Medori, R., Montagna, P., Tritschler, H.J., LeBlanc, A., Cortelli, P., Tinuper, P., Lugaresi, E. & Gambetti, P. 1992*a* Fatal familial insomnia: A second kindred with mutation of prion protein gene at codon 178. *Neurology* **42**, 669–670.

Medori, R., Tritschler, H.-J., LeBlanc, A., Villare, F., Manetto, V., Chen, H.Y., Xue, R., Leal, S., Montagna, P., Cortelli, P., Tinuper, P., Avoni, P., Mochi, M., Baruzzi, A., Hauw, J.J., Ott, J., Lugaresi, E., Autilio-Gambetti, L. & Gambetti, P. 1992*b* Fatal familial insomnia, a prion disease with a mutation at codon 178 of the prion protein gene. *N. Engl. J. Med.* **326**, 444–449.

Meyer, R.K., McKinley, M.P., Bowman, K.A., Braunfeld, M.B., Barry, R.A. & Prusiner, S.B. 1986 Separation and properties of cellular and scrapie prion proteins. *Proc. natn. Acad. Sci. U.S.A.* **83**, 2310–2314.

Meyer, N., Rosenbaum, V., Schmidt, B., Gilles, K., Mirenda, C., Groth, D., Prusiner, S.B. & Riesner, D. 1991 Search for a putative scrapie genome in purified prion fractions reveals a paucity of nucleic acids. *J. Gen. Virol.* **72**, 37–49.

Milson, G., Hunter, G.D. & Kimberlin, R.H. 1971 An experimental examination of the scrapie agent in cell membrane mixtures. II. The association of scrapie infectivity with membrane fractions. *J. comp. Pathol.* **81**, 255–265.

Mobley, W.C., Neve, R.L., Prusiner, S.B. & McKinley, M.P. 1988 Nerve growth factor increases mRNA levels for the prion protein and the beta-amyloid protein precursor in developing hamster brain. *Proc. natn. Acad. Sci. U.S.A.* **85**, 9811–9815.

Murdoch, G.H., Sklaviadis, T., Manuelidis, E.E. & Manuelidis, L. 1990 Potential retroviral RNAs in Creutzfeldt–Jakob disease. *J. Virol.* **64**, 1477–1486.

Neary, K., Caughey, B., Ernst, D., Race, R.E. & Chesebro, B. 1991 Protease sensitivity and nuclease resistance of the scrapie agent propagated in vitro in neuroblastoma-cells. *J. Virol.* **65**, 1031–1034.

Neugut, R.H., Neugut, A.I., Kahana, E., Stein, Z. & Alter, M. 1979 Creutzfeldt–Jakob disease: familial clustering among Libyan-born Israelis. *Neurology* **29**, 225–231.

Nochlin, D., Sumi, S.M., Bird, T.D., Snow, A.D., Leventhal, C.M., Beyreuther, K. & Masters, C.L. 1989 Familial dementia with PrP-positive amyloid plaques: a variant of Gerstmann–Sträussler syndrome. *Neurology* **39**, 910–918.

Oesch, B., Westaway, D., Wälchli, M., McKinley, M.P., Kent, S.B.H., Aebersold, R., Barry, R.A., Tempst, P., Teplow, D.B., Hood, L.E., Prusiner, S.B. & Weissmann, C. 1985 A cellular gene encodes scrapie PrP 27-30 protein. *Cell* **40**, 735–746.

Oesch, B., Groth, D.F., Prusiner, S.B. & Weissmann, C. 1988 Search for a scrapie-specific nucleic acid: a progress report. In *Novel infectious agents and the central nervous system, Ciba Foundation Symposium 135* (ed. G. Bock & J. Marsh), pp. 209–223. Chichester: John Wiley and Sons.

Owen, F., Poulter, M., Lofthouse, R., Collinge, J., Crow, T.J., Risby, D., Baker, H.F., Ridley, R.M., Hsiao, K. & Prusiner, S.B. 1989 Insertion in prion protein gene in familial Creutzfeldt–Jakob disease. *Lancet* (i), 51–52.

Owen, F., Poulter, M., Collinge, J. & Crow, T.J. 1990*a* Codon 129 changes in the prion protein gene in Caucasians. *Am. J. hum. Genet.* **46**, 1215–1216.

Owen, F., Poulter, M., Shah, T., Collinge, J., Lofthouse, R., Baker, H., Ridley, R., McVey, J. & Crow, T. 1990*b* An in-frame insertion in the prion protein gene in familial Creutzfeldt–Jakob disease. *Molec. Brain Res.* **7**, 273–276.

Owen, F., Poulter, M., Collinge, J., Leach, M., Shah, T., Lofthouse, R., Chen, Y.F., Crow, T.J., Harding, A.E. & Hardy, J. 1991 Insertions in the prion protein gene in atypical dementias. *Expl Neurol.* **112**, 240–242.

Owen, F., Poulter, M., Collinge, J., Leach, M., Lofthouse, R., Crow, T.J. & Harding, A.E. 1992 A dementing illness associated with a novel insertion in the prion protein gene. *Molec. Brain Res.* **13**, 155–157.

Palmer, M.S., Dryden, A.J., Hughes, J.T. & Collinge, J. 1991 Homozygous prion protein genotype predisposes to sporadic Creutzfeldt–Jakob disease. *Nature, Lond.* **352**, 340–342.

Parry, H.B. 1962 Scrapie: a transmissible and hereditary disease of sheep. *Heredity* **17**, 75–105.

Parry, H.B. 1983 *Scrapie disease in sheep.* (192 pages.) New York: Academic Press.

Pattison, I.H. 1965 Experiments with scrapie with special reference to the nature of the agent and the pathology of the disease. In *Slow, latent and temperate virus infections, NINDB monograph 2* (ed. D. C. Gajdusek, C. J. Gibbs Jr & M. P. Alpers), pp. 249–257. Washington, D.C.: U.S. Government Printing.

Pattison, I.H. 1966 The relative susceptibility of sheep, goats and mice to two types of the goat scrapie agent. *Res. vet. Sci.* **7**, 207–212.

Pattison, I.H. 1988 Fifty years with scrapie: a personal reminiscence. *Vet. Rec.* **123**, 661–666.

Pattison, I.H. & Jones, K.M. 1967 The possible nature of the transmissible agent of scrapie. *Vet. Rec.* **80**, 1–8.

Prusiner, S.B. 1982 Novel proteinaceous infectious particles cause scrapie. *Science, Wash.* **216**, 136–144.

Prusiner, S.B. 1991 Molecular biology of prion diseases. *Science, Wash.* **252**, 1515–1522.

Prusiner, S.B., Hadlow, W.J., Garfin, D.E., Cochran, S.P., Baringer, J.R., Race, R.E. & Eklund, C.M. 1978 Partial purification and evidence for multiple molecular forms of the scrapie agent. *Biochemistry* **17**, 4993–4997.

Prusiner, S.B., Groth, D.F., Cochran, S.P., Masiarz, F.R., McKinley, M.P. & Martinez, H.M. 1980 Molecular properties, partial purification, and assay by incubation period measurements of the hamster scrapie agent. *Biochemistry* **19**, 4883–4891.

Prusiner, S.B., McKinley, M.P., Groth, D.F., Bowman, K.A., Mock, N.I., Cochran, S.P. & Masiarz, F.R. 1981 Scrapie agent contains a hydrophobic protein. *Proc. natn. Acad. Sci. U.S.A.* **78**, 6675–6679.

Prusiner, S.B., Bolton, D.C., Groth, D.F., Bowman, K.A., Cochran, S.P. & McKinley, M.P. 1982*a* Further purification and characterization of scrapie prions. *Biochemistry* **21**, 6942–6950.

Prusiner, S.B., Gazdusek, D.C., Alpers, M.P. 1982*b* Kuru with incubation periods exceeding two decades. *Ann. Neurol.* **12**, 1–9.

Prusiner, S.B., McKinley, M.P., Bowman, K.A., Bolton, D.C., Bendheim, P.E., Groth, D.F. & Glenner, G.G. 1983 Scrapie prions aggregate to form amyloid-like birefringent rods. *Cell* **35**, 349–358.

Prusiner, S.B., Groth, D.F., Bolton, D.C., Kent, S.B. & Hood, L.E. 1984 Purification and structural studies of a major scrapie prion protein. *Cell* **38**, 127–134.

Prusiner, S.B., Scott, M., Foster, D., Pan, K.-M., Groth, D., Mirenda, C., Torchia, M., Yang, S.-L., Serban, D., Carlson, G.A., Hoppe, P.C., Westaway, D. & DeArmond, S.J. 1990 Transgenetic studies implicate interactions between homologous PrP isoforms in scrapie prion replication. *Cell* **63**, 673–686.

Puckett, C., Concannon, P., Casey, C. & Hood, L. 1991 Genomic structure of the human prion protein gene. *Am. J. hum. Genet.* **49**, 320–329.

Race, R.E., Graham, K., Ernst, D., Caughey, B. & Chesebro, B. 1990 Analysis of linkage between scrapie incubation period and the prion protein gene in mice. *J. gen. Virol.* **71**, 493–497.

Raeber, A.J., Borchelt, D.R., Scott, M. & Prusiner, S.B. 1992 Attempts to convert the cellular prion protein into the scrapie isoform in cell-free systems. *J. Virol.* **66**, 6155–6163.

Richardson, E.P.J. 1977 Myoclonic dementia – Introduction. In *Neurological classics in modern translation* (ed. D. A. Rottenberg & F. H. Hochberg), pp. 95–96. New York: Hafner Press.

Rogers, M., Taraboulos, A., Scott, M., Groth, D. & Prusiner, S.B. 1990 Intracellular accumulation of the cellular prion protein after mutagenesis of its Asn-linked glycosylation sites. *Glycobiology* **1**, 101–109.

Safar, J., Ceroni, M., Piccardo, P., Liberski, P.P., Miyazaki, M., Gajdusek, D.C. & Gibbs, C.J. Jr 1990 Subcellular distribution and physicochemical properties of scrapie associated precursor protein and relationship with scrapie agent. *Neurology* **40**, 503–508.

Scott, M., Foster, D., Mirenda, C., Serban, D., Coufal, F., Wälchli, M., Torchia, M., Groth, D., Carlson, G., DeArmond, S.J., Westaway, D. & Prusiner, S.B. 1989 Transgenic mice expressing hamster prion protein produce species-specific scrapie infectivity and amyloid plaques. *Cell* **59**, 847–857.

Stahl, N., Borchelt, D.R., Hsiao, K. & Prusiner, S.B. 1987 Scrapie prion protein contains a phosphatidylinositol glycolipid. *Cell* **51**, 229–240.

Stahl, N., Baldwin, M.A., Burlingame, A.L. & Prusiner, S.B. 1990*a* Identification of glycoinositol phospholipid-linked and truncated forms of the scrapie prion protein. *Biochemistry* **29**, 8879–8884.

Stahl, N., Borchelt, D.R. & Prusiner, S.B. 1990*b* Differential release of cellular and scrapie prion proteins from cellular membranes by phosphatidylinositol-specific phospholipase C. *Biochemistry* **29**, 5405–5412.

Stahl, N., Baldwin, M.A., Hecker, R., Pan, K.-M., Berlingame, A.L. & Prusiner, S.B. 1992*a* Glycosylinositol phospholipid anchors of the scrapie and cellular prion proteins contain sialic acid. *Biochemistry* **31**, 5043–5053.

Stahl, N., Baldwin, M.A., Teplow, D., Hood, L.E., Beavis, R., Chait, B., Gibson, B., Burlingame, A.L. & Prusiner, S.B. 1992*b* Cataloguing post-translational modifications of the scrapie prion protein by mass spectrometry. In *Prion diseases of humans and animals* (ed. S. B. Prusiner, J. Collinge, J. Powell & B. Anderton), pp. 361–379. London: Ellis Horwood.

Tagliavini, F., Prelli, F., Ghisto, J., Bugiani, O., Serban, D., Prusiner, S.B., Farlow, M.R., Ghetti, B. & Frangione, B. 1991 Amyloid protein of Gerstmann–Sträussler–Scheinker disease (Indiana kindred) is an 11-kd fragment of prion protein with an N-terminal glycine at codon 58. *EMBO J.* **10**, 513–519.

Taraboulos, A., Rogers, M., Borchelt, D.R., McKinley, M.P., Scott, M., Serban, D. & Prusiner, S.B. 1990*a* Acquisition of protease resistance by prion proteins in scrapie-infected cells does not require asparagine-linked glycosylation. *Proc. natn. Acad. Sci. U.S.A.* **87**, 8262–8266.

Taraboulos, A., Serban, D. & Prusiner, S.B. 1990*b* Scrapie prion proteins accumulate in the cytoplasm of persistently-infected cultured cells. *J. Cell Biol.* **110**, 2117–2132.

Taraboulos, A., Jendroska, K., Serban, D., Yang, S.-L., DeArmond, S.J. & Prusiner, S.B. 1992*a* Regional mapping of prion proteins in brains. *Proc. natn. Acad. Sci. U.S.A.* **89**, 7620–7624.

Taraboulos, A., Raeber, A., Borchelt, D.R., Serban, D. & Prusiner, S.B. 1992*b* Synthesis and trafficking of prion proteins in cultured cells. *Molec. Biol. Cell.* **3**, 851–863.

Tateishi, J., Kitamoto, T., Doh-ura, K., Sakaki, Y., Steinmetz, G., Tranchant, C., Warter, J.M. & Heldt, N. 1990 Immunochemical, molecular genetic, and transmission studies on a case of Gerstmann–Sträussler–Scheinker syndrome. *Neurology* **40**, 1578–1581.

Udovitch, A.L. & Valensi, L. 1984 The last Arab Jews: the communities of Jerba, Tunisia. (178 pages.) London: Harwood Academic Publishers.

Vnencak-Jones, C.L. & Phillips, J.A. 1992 Identification of heterogeneous PrP gene deletions in controls by detection of allele-specific heteroduplexes (DASH). *Am. J. hum. Genet.* **50**, 871–872.

Weissmann, C. 1991*a* A "unified theory" of prion propagation. *Nature, Lond.* **352**, 679–683.

Weissmann, C. 1991*b* Spongiform encephalopathies – the prion's progress. *Nature, Lond.* **349**, 569–571.

Westaway, D., Goodman, P.A., Mirenda, C.A., McKinley, M.P., Carlson, G.A. & Prusiner, S.B. 1987 Distinct prion proteins in short and long scrapie incubation period mice. *Cell* **51**, 651–662.

Westaway, D., Mirenda, C.A., Foster, D., Zebarjadian, Y., Scott, M., Torchia, M., Yang, S.-L., Serban, H., DeArmond, S.J., Ebeling, C., Prusiner, S.B. & Carlson, G.A. 1991 Paradoxical shortening of scrapie incubation times by expression of prion protein transgenes derived from long incubation period mice. *Neuron* **7**, 59–68.

Wexler, N.S., Young, A.B., Tanzi, R.E., Travers, H.,

Starosta-Rubinstein, S., Penney, J.B., Snodgrass, S.R., Shoulson, I., Gomez, F., Ramos Arroyo, M.A., Penchaszadeh, G.K., Moreno, H., Gibbons, K., Faryniarz, A., Hobbs, W., Anderson, M.A., Bonilla, E., Conneally, P.M. & Gusella, J.F. 1987 Homozygotes for Huntington's disease. *Nature, Lond.* **326**, 194–197.

Wilesmith, J. & Wells, G.A.H. 1991 Bovine spongiform encephalopathy. *Curr. Top. Microbiol. Immunol.* **172**, 21–38.

Wilesmith, J.W., Wells, G.A.H., Cranwell, M.P. & Ryan, J.B.M. 1988 Bovine spongiform encephalopathy: epidemiological studies. *Vet. Rec.* **123**, 638–644.

Wilesmith, J.W., Hoinville, L.J., Ryan, J.B.M. & Sayers, A.R. 1992*a* Bovine spongiform encephalopathy: aspects of the clinical picture and analyses of possible changes 1986–1990. *Vet. Rec.* **130**, 197–201.

Wilesmith, J.W., Ryan, J.B.M., Hueston, W.D. & Hoinville, L.J. 1992*b* Bovine spongiform encephalopathy: epidemiological features 1985 to 1990. *Vet. Rec.* **130**, 90–94.

Yost, C.S., Lopez, C.D., Prusiner, S.B., Meyers, R.M. & Lingappa, V.R. 1990 A non-hydrophobic extracytoplasmic determinant of stop transfer in the prion protein. *Nature, Lond.* **343**, 669–672.

Index